MAN AND BEAST.

MAN AND BEAST

HERE AND HEREAFTER.

ILLUSTRATED BY MORE THAN THREE HUNDRED
ORIGINAL ANECDOTES.

BY THE

Rev. J. G. WOOD, M.A., F.L.S.,

AUTHOR OF "HOMES WITHOUT HANDS," &c.

"I canna but believe that dowgs hae sowls."
JAMES HOGG, *the Ettrick-Shepherd.*

NEW YORK:
HARPER & BROTHERS, PUBLISHERS,
FRANKLIN SQUARE.
1875.

BY J. G. WOOD.

HOMES WITHOUT HANDS: being a Description of the Habitations of Animals, classed according to their Principle of Construction. By J. G. WOOD, M.A., F.L.S., Author of "Illustrated Natural History." With about 140 Illustrations engraved on Wood by G. PEARSON, from Original Designs made by F. W. KEYL and E. A. SMITH, under the Author's Superintendence. 8vo, Cloth, $4 50; Sheep, $5 00; Half Calf, $6 75.

Mr. Wood's classification of the habitations of animals opens up so wide and connected a perspective into the *psychology* of the animal creation that it possesses quite a peculiar interest and fascination. The facts that rats and mice live in holes, and birds make nests, taken in an isolated way, leaves little impression upon the imagination of the master-builder man, drunk with his own glories, who looks with pity and contempt upon dwellings and contrivances which, if not in grandeur, at all events in delicacy of adaptation and ingenuity of workmanship, rival his own. Niebuhr said somewhere or other that genius is seen in the magnitude of the results compared with the slenderness of the materials attainable. According to this definition, many animals exhibit far more than instinct—they show genius in the construction of their homes. But it is only when we study them on some such connected plan as that furnished by Mr. Wood that we gradually become irresistibly impressed by sheer cumulative force, rather than direct proof, with the absurdity of the popular talk about blind instinct, and can not help seeing the large amount of downright solid intellect which birds and beasts bring to bear on the construction of their homes.—*Spectator*, London.

PUBLISHED BY HARPER & BROTHERS, NEW YORK.

☞ HARPER & BROTHERS *will send the above work by mail, postage prepaid, to any part of the United States or Canada, on receipt of the price.*

PREFACE.

In the opening of Bishop Butler's "Analogy of Religion" the following passage occurs, showing that this eminent divine considered the lower animals as capable of a future life: "It is said these observations are equally applicable to brutes; and it is thought an insuperable difficulty that they should be immortal, and by consequence capable of everlasting happiness. Now this manner of expression is both invidious and weak; but the thing intended by it is really no difficulty at all, either in the way of natural or moral consideration."

The Bishop then refers to the "latent powers and capacities" of the lower animals, and sees no reason why they should not be developed in a future life. In the present work, I have endeavored to carry out his train of thought, and to show that the lower animals do possess those mental and moral characteristics which we admit in ourselves to belong to the immortal spirit, and not to the perishable body.

The scheme of the book is briefly as follows. I begin with clearing away the difficulties which arise from two misunderstood passages in the Old Testament, and prove that the Scriptures do not deny a future life to the lower animals. I then show that the lower animals share with man the attributes of Reason, Language, Memory, a sense of Moral Responsibility, Unselfishness, and Love, all of which belong to the spirit and not to the body; and that as man expects to retain these qualities in the next world, there is every reason to presume that the lower animals may share his immortality hereafter as they share his mortality at present.

In order to prove that animals really possess the above-mentioned qualities, I cite more than three hundred original anecdotes, all being authenticated by the writers, and the documents themselves remaining in my possession.

J. G. W.

CONTENTS.

CHAPTER		PAGE
I.	THE TESTIMONY OF REVELATION	9
II.	REASON AND INSTINCT	16
III.	REASON—(*continued*)	24
IV.	REASON—(*concluded*)	32
V.	LANGUAGE [OF ANIMALS]	39
VI.	LANGUAGE [HUMAN]	50
VII.	MEMORY	60
VIII.	GENEROSITY	68
IX.	CHEATERY	75
X.	HUMOR	79
XI.	PRIDE, JEALOUSY, ANGER, REVENGE, TYRANNY	89
XII.	CONSCIENCE	97
XIII.	SYMPATHY AND FRIENDSHIP	104
XIV.	LOVE OF MASTER	114
XV.	CONJUGAL LOVE	126
XVI.	PARENTAL LOVE	128
XVII.	THE FUTURE STATE	136

MAN AND BEAST.

CHAPTER I.

THE TESTIMONY OF REVELATION.

The Future of the Lower Animals, as popularly supposed to be Taught in the Scriptures.—The "Beasts that perish."—If the Literal Sense of the Scriptures alone be taken, the Future Life of Man is repeatedly denied in the Books of Psalms, Job, and Ecclesiastes.—The Necyomanteia of Homer compared with the Psalms and Ecclesiastes.—The Future State of Man according to Horace.—Comparison of the Renderings of Psa. xlix. 20 (the "beasts that perish") in the Hebrew, Greek, Latin, English (Douay version), German, Spanish, Italian, French, Chaldaic, Syriac, and Arabic Versions.—Subject of Psa. xlix., and the Real Signification of the Concluding Verse.—Opinions of Correspondents.—The "Spirit of the Beast that goeth downward to the Earth."—Subject of the Book called Ecclesiastes.—Teaching by means of Irony.—Distinction between the Spirit of Man and that of the Lower Animals.

IN dealing with a subject of this nature—namely, the spiritual condition of the animals inferior to Man—it is clear that we must, in the first place, refer to the Scriptures, from which is derived all our authentic knowledge of spiritual life.

There is a popular belief—I should rather say a popular tradition — that somewhere in the Scriptures we are taught that, of all living inhabitants of earth, Man alone possesses a spirit, and that therefore he alone survives in spirit after the death of the material body. If this were true, there would be no room for argument to those who profess to believe the Scriptures literally, and to base their faith upon that literal belief; and, however such a statement might seem to controvert all ideas of benevolence, justice, and even common-sense, such believers would be bound to receive it on trust, and to wait for a future time in which to understand it.

Many persons go so far as to deny to animals even the possession of Reason, and only attribute to them the power of Instinct, while there are comparatively few who do not believe that when an animal dies, its life-principle dies too—that the animating power is annihilated, while the body is resolved into its various elements so as to take form in other bodies.

This belief is almost entirely, if not wholly, due to two passages of Scripture, one being in the Psalms, and the other in Ecclesiastes. The former is that which is generally quoted as decisive of the whole question. It runs in the authorized version as follows : "Nevertheless, man being in honor, abideth not; he is like the beasts that perish" (Psa. xlix. 12, 20).

The Prayer-book version is somewhat different, but is yet translated much to the same effect. "Man, being in honor, hath no understanding, but is compared to the beasts that perish."

The second passage occurs in Ecclesiastes iii. 21: "Who knoweth the spirit of man that goeth upward, and the spirit of the beast that goeth downward to the earth."

On the strength of these two passages, we are called upon to believe that when a beast dies, it dies forever, and that its life is utterly extinguished as is the flame of an expired lamp. Now every one who has had even a slight acquaintance with the exposition of Scripture is aware that nothing is more dangerous than attempting to explain any passage, however simple it may appear to be, without making a reference to the original text. The translator may have mistaken the true sense of the words; or he may have insufficiently expressed their signification; or, owing to a change in the meaning of words, a passage may now bear on its surface an exactly contrary sense to that which it conveyed when it was first written.

However, we will lay aside that point for the

present, and accept the passage as it stands, together with the literal signification of the words as generally understood.

There will then be no doubt that we must believe that beasts have no immortal life. But, if we are to take the literal sense of the Bible, and no other, we are equally bound to believe that Man as well as beast has no life after death.

See, for example, Psa. vi. 5: "In death there is no remembrance of thee: in the grave, who shall give thee thanks?"

Also, Psa. lxxxviii. 10, 11, 12:

"Wilt thou show wonders to the dead? Shall the dead arise and praise thee?

"Shall thy loving-kindness be declared in the grave, or thy faithfulness in destruction?

"Shall thy wonders be known in the dark, and thy righteousness in the land of forgetfulness?"

Also, see Psa. cxv. 17: "The dead praise not the Lord, neither any that go down into silence."

Also, Psa. cxliii. 3: "For the enemy hath persecuted my soul; he hath smitten my life down to the ground; he hath made me to dwell in darkness, as those that have been long dead."

Also, Psa. cxlvi. 3, 4:

"Put not your trust in princes, nor in the son of man, in whom there is no help.

"His breath goeth forth, he returneth to his earth; in that very day his thoughts perish."

If we are to take the Scriptures solely in their literal sense, there can be no doubt of their meaning. The whole range of heathen literature contains nothing more gloomy, dreary, or more despondent in the contemplation of death. "Let us eat and drink, for to-morrow we die," would be a fit result of such a belief.

In the very book in which occurs the single passage on which is based the denial of the immortality of the lower animals are five passages which proclaim the same end to the life of man. We are told distinctly and definitely that those who have died have no remembrance of God, and can not praise him. Death is described as the "land of forgetfulness"—the place of darkness, where all man's thoughts perish. Can more than this be said of the "beasts that perish?"

Now we will leave the Psalmist, and proceed to other writers. Treating, not of the wicked, but of mankind in general who "dwell in houses of clay," the writer proceeds as follows: "They are destroyed from morning to evening; *they perish forever*, without any regarding it" (Job iv. 20).

Take another passage from the same book, a passage which is even more definite in its statement: "As the cloud is consumed and vanisheth away, so he that goeth down to the grave shall come up no more" (Job vii. 9).

Again—

"Man dieth, and wasteth away: yea, man giveth up the ghost, and where is he?

"As the waters fail from the sea, and the flood decayeth and drieth up:

"So man lieth down, and riseth not" (Job xiv. 10, 11, 12). And ver. 14: "If a man die, shall he live again?"

See also the piteous wail of Job over his life as shown in chap. iii. and x. In the first he complains that he was ever born, that being was ever given to him, that he was ever taken out of a state of absolute nonentity. In the second he repeats the same lamentation, with the addition that even death can bring no relief to his sufferings except extinction.

"Wherefore, then, hast thou brought me forth out of the womb? Oh that I had given up the ghost, and no eye had seen me!

"I should have been as though I had not been; I should have been carried from the womb to the grave.

"Are not my days few? Cease then, and let me alone, that I may take comfort a little,

"Before I go whence I shall not return, even to the land of darkness and the shadow of death;

"A land of darkness, as darkness itself; and of the shadow of death, without any order, and where the light is as darkness" (Job x. 18-22).

Turning to the Book of Ecclesiastes, in which occurs the solitary passage which is held to disprove the immortality of the lower animals, we find the following passages, which are even more emphatic as to the future state of man:

"I said in my heart concerning the estate of the sons of men, that God might manifest them, and that they might see that they themselves are beasts.

"For that which befalleth the sons of men befalleth beasts; even one thing befalleth them. As the one dieth, so dieth the other; yea, they have all one breath, so that a man has no preeminence over a beast: for all is vanity.

"All go unto one place; all are of the dust, and all turn to dust again" (Eccles. iii. 18, 19, 20).

Also in ch. ix. 5: "For the living know that they shall die, but the dead know not any thing, neither have they any more a reward, for the memory of them is forgotten."

Also in ch. ix. 10: "Whatsoever thy hand findeth to do, do it with thy might; for there is no work, nor device, nor knowledge, nor wisdom in the grave whither thou goest."

Taking the literal sense of these words and no

other, it is impossible to doubt their import. They state definitely that, as regards a spiritual life, there is no distinction between man and beast; and that when they die, all go to the same place. The writer also distinctly states that after death man can work nothing, know nothing, nor can receive any reward. The same vein of irrepressible sadness that characterizes the extracts taken from the Psalms is prominent in those passages from Job and Ecclesiastes; and if from these alone we were to deduce our ideas of the condition of man after death, most sad and hopeless would be the very thought of dissolution.

It is true that we do not accept them in this light, knowing that they are written symbolically or parabolically, and that there underlies them the spiritual sense of which St. Paul speaks when he contrasts the life-giving spirit with the death-dealing letter (2 Cor. iv. 6). With that meaning, however, we have in the present case nothing to do. We are only concerned with the literal meaning of our translation, and, according to that literal meaning, if we take two texts to prove that beasts have no future life, we are forced by no less than fourteen passages to believe that Man, in common with beasts, has no future life. We have no right to pick and choose which passages we are to take literally, and which symbolically, but must apply the same test to all alike, and treat all in the same manner.

Let us pass for a while from sacred to secular literature. All my classical readers must be familiar with that wonderful eleventh book of Homer's "Odyssey" generally called the Necyomanteia, or Invocation of the Dead. In this strange history Ulysses is shown as descending into the regions inhabited by departed spirits, for the purpose of invoking them and obtaining their advice as to his future adventures.

He sails to the boundaries of the ocean, and lands in the country of the Cimmerians, who dwell in perpetual cloud and darkness, and in whose country are the gates leading to the regions of the dead. He utters solemn prayers and invocations, offers sacrifices, and pours their blood into a trench of a cubit square, which had been consecrated for that purpose. Straightway there throng around the trench the spirits of the dead, eager to drink the blood, and so to be able to hold converse with one who was still a denizen of the upper world. See Pope's version of the passage:

"Thus solemn rites and holy vows we paid
To all the phantom nations of the dead.
Then died the sheep; a purple torrent flowed,
And all the cavern smoked with streaming blood.
When lo! appeared along the dusky coasts
Thin, airy shoals of visionary ghosts.
Fair pensive youths and soft enamored maids,
And withered elders, pale and wrinkled shades:
Ghastly with wounds, the forms of warriors slain
Stalked with majestic port, a martial train:
These and a thousand more swarmed o'er the ground,
And all the dire assembly shrieked around."

The hero stands over the trench, defending it with his sword from the hosts of the dead, and only allowing the spirits to drink the blood one by one. Thus he converses with the spirits of his companions Elpenor and Tiresias, then sees his mother Anticlea; and at last the spirit of Achilles approaches. The dialogue between the inhabitant of the earth and the denizen of the regions of the dead must be quoted entire:

"Through the thick gloom his friend Achilles knew,
And as he speaks the tears dissolve in dew.
'Comest thou alive to view the Stygian bounds,
Where the wan spectres walk eternal rounds;
Nor fear'st the dark and dismal waste to tread,
Thronged with pale ghosts familiar with the dead?'
To whom with sighs, 'I pass these dreadful gates
To seek the Theban, and consult the Fates;
For still distressed I rove from coast to coast,
Lost to my friends and to my country lost.
But sure the eye of Time beholds no name
So blessed as thine in all the rolls of fame:
Alive we hailed thee with our guardian gods,
And, dead, thou rulest a king in these abodes.'
'Talk not of ruling in this dolorous gloom,
Nor think vain words (he cried) can ease my doom.
Rather I'd choose laboriously to bear
A weight of woes and breathe the vital air,
A slave to some poor hind that toils for bread,
Than reign the sceptred monarch of the dead."

Coleridge well remarks of this passage, and indeed of the whole of the Necyomanteia, that it is "remarkable for the dreary and even horrible revelations which it makes of the condition of the future life. All is wild and dark; hunger and thirst and discontent prevail. We hear nothing of elysian fields for piety or wisdom or valor, and there is something quite deadening in the answer of the shade of Achilles to the consolation of Ulysses."

Gloom, misery, and vain regrets for earth pervade the whole of this episode:

"Now, without number, ghost by ghost arose,
All wailing with unutterable woes.
* * * * * * *
But swarms of spectres rose from deepest hell
With bloodless visage and with hideous yell.
They scream, they shriek; and groans and dismal sounds
Stun my scared ears, and pierce hell's utmost bounds.
No more my heart the dismal din sustains,
And my cold blood hangs shivering in my veins."

These are the ideas of a heathen poet concerning the future state of man. It is no wonder that sensual pleasures should be held the chief object in the life of man, when he is to look forward to such a future as this—a future from which neither wisdom nor virtue nor piety could save him—an eternity of gloom, darkness, repining, and hopeless despondence.

Yet, sad as is this picture of the heathen poet, it is far brighter than that of the Psalmist, the Preacher, or Job.

Those who have passed into the world of spirits do not at all events forfeit their individuality by death. The youth, the maiden, the elder, and the matron are distinguished in the spirit as they had been in the flesh; and those who had lost their lives in honorable battle retain the stern port and martial demeanor of the earthly warrior.

Memory is still left to the dead. They remember their earthly career; they do not lose their interest in their friends who still remain on earth; and, above all, Love survives. Anticlea retains her maternal love for Ulysses, for loss of whom she died; and she watches over the welfare of Penelope and Telemachus. The spirits hold converse with each other. Those who have been friends on the upper earth resume their friendship in the lower regions. Haughty, self-willed, discontented in death as in life—"*Impiger, iracundus, inexorabilis, acer*"—Achilles still receives some solace in the constant companionship of his friend Patroclus.

But, if we are to take literally the passages of Scripture which have been quoted, no such consolation exists in the future state of man, who passes at death into a place of darkness, forgetfulness, and silence, where is no work, nor device, nor knowledge, nor wisdom—where even his very thoughts perish. If these passages are to be understood in their pure literal sense, there is no other interpretation to be put upon them; for the statements are too explicit to be explained away or even softened.

According to the outward sense of their writings, the Psalmist, Job, and the Preacher are very much on a par with Horace in their absolute unbelief in a future existence, and the vein of melancholy which in consequence underlies their utterances. Take, for example, Whittier's short and brilliant analysis of the philosophy of Horace as supposed to be spoken by a friend:

"Speaking of Horace, he gives us glowing descriptions of his winter circles of friends, where mirth and wine, music and beauty, charm away the hours, and of summer-day recreations beneath the vine-wedded elms of the Tiber, or on the breezy slopes of Soracte; yet I seldom read them without a feeling of sadness.

"A low wail of inappeasable sorrow, an undertone of dirges, mingles with his gay melodies. His immediate horizon is bright with sunshine; but beyond there is a world of darkness, the light whereof is darkness. It is walled about by the everlasting night. The skeleton sits at his table; a shadow of the inevitable terror rests upon all his pleasant pictures. He was without God in the world; he had no clear abiding hope of a life beyond that which was hastening to a close. Eat and drink, he tells us; enjoy present health and competence; alleviate present evils, or forget them, in social intercourse, in wine, music, and sensual indulgence; for to-morrow we must die. Death was in his view no mere change of condition and relation; it was the black end of all.

"It is evident that he placed no reliance on the mythology of his time, and that he regarded the fables of the Elysian Fields, and their dim and wandering ghosts, simply in the light of convenient poetic fictions for illustrations and imagery.

"Nothing can, in my view, be sadder than his attempts at consolation for the loss of friends. Witness his Ode to Virgil on the death of Quintilius. He tells his illustrious friend simply that his calamity is without hope, irretrievable and eternal; that it is idle to implore the gods to restore the dead; and that, although his lyre may be more sweet than that of Orpheus, he can not reanimate the shadow of his friend, nor persuade the 'ghost-compelling god' to unbar the gates of death. He urges patience as the sole resource. He alludes not unfrequently to his own death in the same despairing tone.

"In the Ode to Torquatus—one of the most beautiful and touching of all he has written—he sets before his friend, in melancholy contrast, the return of the seasons, and of the moon renewed in brightness, with the end of man, who sinks into the endless dark, leaving nothing behind save ashes and shadows. He then, in the true spirit of his philosophy, urges Torquatus to give his present hour and wealth to pleasures and delights, as he had no assurance of to-morrow."

Compare this analysis with that of the Psalmist, Job, and the Preacher, and the result will be found to be the same in all the cases—namely, an inability to believe in a future life, and a consequent desire to snatch what fleeting pleasures the world can give, before the inevitable Fates consign him to dark oblivion.

It may seem rather startling to compare the teachings of a Greek idolatrous heathen and of a Latin Epicurean heathen with those of sacred writers. Still more startling is it to show that the teachings of the Epicurean sensualist are no worse than those of the Scriptural writer, while those of the Greek poet are very much better. It is, however, the fact, and, if we are to be bound by the literal meaning of the Scriptures, there is no possibility of denying it without doing violence to reason and ordinary common-sense.

Now, however, we come to the point which was mentioned on page 9. Does the authorized version give a full and correct interpretation of the Hebrew text? It certainly does not. There is no change in the significance of the words, there is no mere insufficiency in the translation, but the rendering is absolutely and entirely wrong. The word "perish" does not occur at all in the Hebrew text, nor is even the idea expressed. The words which our translation twice renders as "beasts that perish" are in the Hebrew כַּבְּהֵמוֹת נִדְמוּ, *i. e.*, "dumb beasts." On comparing a number of translations of Psalm xlix. into various languages, I find that scarcely any of them even imply the idea of perishing in the sense of annihilation. First, we will take the "Jewish Bible," which is acknowledged to be the best and closest translation in our language, and which has been made by Dr. Benisch, under the supervision of the Chief Rabbi. Both in verses 12 and 20 the translation is as follows:

"Man *that is* in honor, and understandeth *this* not, is like the beasts *that are* irrational." A foot-note gives the word "dumb," as an alternative reading for "irrational."

The Septuagint has very much the same reading, the verse ending with these words, "παρασυνεβλήθη τοῖς κτήνεσι τοῖς ἀνοήτοις." This is the Vatican text. Sir Lancelot C. Lee Brunton's translation of the Septuagint runs as follows: "Man that is in honor understands not; he is compared to the senseless cattle, and is like them."

Here is the Vulgate:

"Comparatus est jumentis insipientibus, et similis factus est illis."

In Wycliffe's Bible, which is a translation from the Vulgate, the passage is thus rendered:

"A man whanne he was in honour understood not; he is comparisound to unwise beestis, and is maad lijk to tho."

The "Douay" Bible, *i. e.*, the translation of the English Roman Catholic College of Douay, being the version which is accepted by that branch of the Church in England, renders the passage as follows:

"Man, when he was in honor, did not understand; he hath been compared to senseless beasts, and made like to them."

The Ethiopian version, as read by means of a Latin translation, is nearly the same as the Vulgate.

The French and Italian are the only two which resemble our version. The former runs thus:

"L'homme *qui est* en honneur, et *qui* n'a point d'intelligence, est semblable aux bêtes *qui* périssent."

The Italian is as follows:

"L'uomo *che è* in instato onorevole, è e non ha intelletto, simile alle bestie che periscono."

There is a curious Chaldaic version of the passage, which, according to a Latin translation, adds a few words by way of explanation, and, in these words, places *wicked* men and beasts on the same level of nothingness after death. I have placed the additions in brackets:

"Homo [sceleratus] in tempore quo subsistet *in honore*, non intelligit; cum removetur gloria ejus *ab eo*, comparatur bestiæ [et redigitur in nihilum]."

Into some other translations a new idea is imported. Take, for example, Luther's Bible:

"Kurz, wenn ein Mensch in der Würde ist, und hat keinen Verstand, so fähret er davon, wie ein Vieh."

So the Spanish:

"El hombre quando esteba en honor, no lo intendio; ha sido comparado á las bestias insensatas, y se ha hecho semejante á ellas."

The Arabic is almost exactly the same as the Spanish, but ends with the word Alleluia, which is not in the Hebrew.

The Syriac version, according to the Latin translation, conveys a similar idea:

"Homo gloriam suam non intellexit, sed æquavit se animanti et similis factus est ei."

Even supposing that the word "perish" is rendered correctly, it does not follow that annihilation is signified. Take, for example, the tenth verse of the same Psalm in the same version:

"For he seeth that wise men die, and likewise the fool and the brutish person perish, and leave their wealth to others."

Surely no one would interpret this passage as a declaration that the wise and fools and the brutish had no life after the death of the body.

The last verse of the Psalm is, as Luther puts it, a summary of the whole poem. The Psalmist draws a vivid picture of the true object of man's life in this world, and the tendency of man to forget it. He sets forth the shortness of human life, and shows that neither wealth, rank, nor fame can endure after a man dies, all these things belonging to the mere earthly life of man. Consequently, men who set their hearts upon these earthly things ignore the honor of their manhood, and degrade themselves to the level of the dumb beasts, whose aspirations are, as far as we know, limited to this present world.

It will be seen, therefore, that we may dismiss from our minds the idea that the beasts are said by the Psalmist to have no future life, and that we may reject the passage as being totally irrelevant to the subject. It is of the greatest importance that this should be done, as the passage in question is the only one which even appears to make any definite statement as to the future condition of the lower animals.

Some years ago, when writing my "Common Objects of the Country," I ventured to doubt the truth of the popular belief on this subject, and was rather surprised at the result. Almost every periodical which gave a notice of the book quoted the passage, and, with only one or two exceptions, more or less approved of it. The exceptional cases were those of distinctly religious publications, and they of course brought against me "the beasts that perish."

I was also inundated with letters on the subject. Many of them were written by persons who had possessed favorite animals, and who cordially welcomed an idea which they had long held in their hearts, but had been afraid to express. Many were from persons who were seriously shocked at the idea that any animal lower than themselves could live after the death of the body.

Some were full of grave rebuke, while others were couched in sarcastic terms.

Two are specially worthy of notice. The one contains twelve pages of closely written, full-sized letter-paper, in which the writer tells me that any one who cherished the hope that animals could live after death was unworthy of his position of a clergyman, ought to be deprived of his university degrees, and expelled from the learned societies to which he belonged. This argument was so unanswerable that I did not venture to reply to it.

The writer of the second letter remarked that, whatever I might say, he would never condescend to share immortality with a cheese-mite. I replied that, in the first place, it was not likely that he would be consulted on the subject; and that, in the second place, as he did condescend to share mortality with a good many cheese-mites, there could be no great harm in extending his condescension a step further.

But, no matter whether the writers agreed with me or not, no matter whether they were sympathetic, severe, or sarcastic, they invariably mentioned "the beasts that perish." Some wished to know how it was possible to get over a passage which had always prevented them from indulging in the hope that the animals which they had loved on earth would have a future life; while others brought forward "the beasts that perish" as a crushing and conclusive argument, of which they evidently supposed me to be entirely ignorant.

The reader will therefore see how important it is that the true meaning of the Hebrew text should be known, and that the Psalmist should not be accredited with putting forward a doctrine to which, whether true or false, he makes no reference whatever.

Having thus disposed of the "beasts that perish," let us turn to the passage in Ecclesiastes, which, as we have seen, is the only one which has any direct reference to the future state of the lower animals.

"Who knoweth the spirit of man (*or* the sons of man) that goeth upward (*or* ascending), and the spirit of the beast that goeth downward to the earth?" (Eccles. iii. 21).

We have here, at all events, an admission that, whether the spirit ascend or descend, both man and beasts do possess spirits—the Hebrew word being the same in both cases. There is no difference in the various translations, and the rendering in the Jewish Bible is *verbatim* the same as that of our authorized version. We will take the entire passage, and not only an isolated text:

"I said in mine heart concerning the estate of the sons of men, that God might manifest them, and that they might see that they themselves are beasts.

"For that which befalleth the sons of men befalleth beasts; even the one thing befalleth them: as the one dieth, so dieth the other; yea, they have all one breath; so that a man hath no pre-eminence above a beast: for all is vanity.

"All go to one place; all are of the same dust, and all turn to dust again.

"Who knoweth the spirit of man that goeth upward, and the spirit of the beast that goeth downward to the earth?

"Wherefore I perceive that *there is* nothing better than that a man should rejoice in his own works; for that *is* his portion: for who shall bring him to see what shall be after him?" (Eccles. iii. 18 to end of chapter).

The sad, contemptuous irony of the first three chapters of the book tells its own story. Whether or not this book be the production of Solomon in his later years matters very little. It well may be so, for it is the confession of one who has possessed well-nigh all that earth can give him, and who has lived to see its emptiness. Indulgence has been avenged by satiety, and the writer's summary of life is contained in the despondent avowal, "Vanity of vanities, all is vanity."

Self-reproach for a wasted life breathes in every page of this book; and the Preacher, speaking from his own experience, shows that wealth, glory, pleasure, and even wisdom are in themselves but utter emptiness. Practically the theme is the same as that of the forty-ninth Psalm, though the two writers handle it in opposite ways. The Psalmist approaches the subject with grave solemnity, warning his hearers of the brevity of human life, and showing that if man forgets the glory of his manhood, made in the image of God, he places himself on the level of the dumb beasts.

The Preacher takes a different view of the case, though he comes to the same conclusion. Employing biting sarcasm instead of solemn warning, he first shows the utter emptiness of all worldly and selfish pleasures, and the miserable end of the voluptuary, and then ironically advises his readers to place their whole happiness in them.

Briefly, this is the argument: Suppose any one may say that this is living a mere animal life, what of that? Who *could* be expected to know that the spirit of beasts is inferior to that of man, and that the spirit of man was made to soar above earthly things, while that of beasts is limited to them?

The bitter irony is evident, and through the book this idea repeatedly occurs under various forms.

But by no manner of interpretation can the twenty-first verse mean that beasts are annihilated after death, while men rise again. The writer ironically assumes that his readers do not know the difference between the spirit of man and that of beast, and, arguing from that assumption, advises them to live a mere animal life.

"*There is* nothing better for a man *than that* he should eat and drink, and that he should make his soul enjoy good in his labor."

I have already shown that the former of these passages does not even contain the idea of annihilation as regards beasts; and that the latter is entirely misapprehended is now evident. We may therefore dismiss from our minds both Psa. xlix. and Eccles. iii. as having no bearing whatever on the subject. The Scriptures therefore, as far as we have seen, do not deny future life to the lower animals. Whether they assert it, is not relevant to the present issue.

CHAPTER II.

REASON AND INSTINCT.

Distinction between Instinct and Reason.—Definition of Instinct.—Rarey the Horse-tamer.—Various Phases of Instinct in Man and Beast.—Definition of Reason.—Comparison between Children and Animals.—Reasoning Powers of the Fishes and Reptiles.—Reason Displayed by the Common Toad.—The Axolotl and the Horned Toad.—Two "Temperance" Dogs and their Masters.—"Mess" and his Ways.—Knowledge of his Regimental Uniform.—Methodical Habits.—Medicine and Nightcaps.—A Broken Leg and its Consequences.—Unexpected Failure of Reasoning in my Dog "Apollo."

HAVING now disposed of the purely theological objections to the future life of the lower animals, we proceed to the subject which necessarily follows next in order — namely, the possession of reasoning powers in them.

There is much vagueness of idea on this point, the general tendency being to confound reason and instinct together, and to wonder when one ends and the other begins. For example, there are hundreds of anecdotes, too familiar for quotation or even mention, which are described as wonderful examples of instinct, whereas every one of them is a proof of reason, and has nothing to do with instinct.

When the late Mr. Rarey was exhibiting his wonderful powers of horse-taming in England, I had a long argument with him. It was his custom to preface his performances by a short lecture, in which he was in the habit of saying that he conquered the animals because he possessed reason and the horse did not. I submitted to him that his words and his actions were diametrically opposed to each other; for that, while he denied reason in the horse, every successive stage in the education of the animal was a direct appeal to its reason.

His success was really due to the higher and more comprehensive reason subduing the lower and more limited; while, if the horse did not possess reason, Mr. Rarey could have exercised no influence whatever upon it. Indeed, as he had stated in his lecture that dull and stupid horses were more difficult to tame than intelligent and high-spirited animals, he had already granted their capacity of reasoning.

Some years ago I had a standing dispute with my valued friend, the late Charles Waterton. Swayed probably by his religious views, which were of the severest character, he never would admit, and never did admit, that any animal lower than man could possess reason. Yet in all his dealings with the animal world, in which he was simply without a rival, he invariably appealed to their reason and not to their instinct.

For example, he never would allow his farm horses to be tied up or even shut in their stalls after their day's labor. He always had them fed in loose boxes, and the doors left open, so that after their meal the animals could go into the yard and talk to each other. "*We* like to chat over our meals," said the kindly old man, " and so do they."

I remember one case in which the bailiff was within an ace of being dismissed on the spot because one of the men had fastened a horse in its stall. Mr. Waterton quietly took out his penknife, cut the halter into inch lengths, laid them on the floor of the stable, and went his way. At six next morning the bailiff waited on his master, looking very much as if he were going to be hanged with a like halter. He knew perfectly well the sort of reception which he would meet, and was evidently relieved when he left the room and found himself still bailiff.

Whenever Waterton showed himself there was a general commotion in the domain, all the inhabitants recognizing their friend and trying to get near him.

One scene I never shall forget. There was a splendid young bull, milk white, one of the many favorites of its master, and the terror of the farm-laborers. It was a high-spirited and playful young beast, and when let out of the stable used to indulge in pranks that are very becoming to a kitten, but seem rather out of place when performed by a bull.

One morning I accompanied Mr. Waterton through the farm-yard, and he casually ordered the men to let "Tommy" loose. When we came around again Tommy was still in his stall, not one of the men having dared to touch him. His master, after calling the men a set of cowards, proceeded to loose Tommy himself, whereupon the men all armed themselves with big cudgels. These Mr. Waterton immediately took away again, just as he removed the weapons of his assistants when he captured the huge snake in Guiana, telling them that if they were afraid they might go; which they did.

He then loosed Tommy, who came plunging out in the exuberance of his freedom, capering about, first on his fore and then on his hind legs, and tossing bundles of litter into the air with his horns. Once he rushed at the great feeding-crib that stood in the middle of the yard, knocked it down, picked it up with his horns, and was on the point of smashing it to pieces, when the men, who were in a horrible fright on the other side of the fence, threw ropes over it and dragged it out of the animal's reach.

Tommy then made a charge at Mr. Waterton, and came straight on him with head down as if he meant to fling him after the crib. I felt rather nervous myself at this; but Mr. Waterton took it with perfect coolness, and just put his hand on the beast's broad white forehead, saying in a tone of mild remonstrance, "Now then, old boy!" Whereupon Tommy kicked up his heels, gave himself a shake, and next moment was prancing all over the yard.

There was not the least harm about the animal. He only wanted to let off the superfluous steam, so to speak, and indulged himself in the absurd antics which have been described. It is very likely that if he saw the men running away he would run after them, thinking that they were joining in his game, whereas they thought that he was going to immolate them on the spot.

In a few minutes Tommy's exuberant spirits had calmed down, and he was seen quietly lying down in the straw with his master seated on him, and feeding him with bits of bread out of his storehouse of a pocket.

I should have liked to have painted that scene: the great white bull lying on the ground with the tall, spare form of his master seated on its huge body; the litter all tossed here and there by his pranks; the horses, cows, cats, poultry, and all sorts of birds crowding around in hope of scraps; and in the background the shamefaced laborers, still in undisguised terror of the bull, and in equally undisguised admiration of their master's courage.

The reader will see here that Mr. Waterton was simply gauging the reason of the bull by means of his own, and that, if the animal had only possessed instinct instead of reason, his master would in all probability have been killed on the spot.

B

Let us define clearly the distinction between instinct and reason.

The well-known and perfectly correct definition of Instinct is this—"*A certain power or disposition of mind by which, independent of all instruction or experience, animals are directed unerringly to do spontaneously whatever is necessary for the preservation of the individual or the continuation of the species.*"

Take ourselves, for example. It is instinct which teaches the child to seek its mother's breast and to obtain its nourishment by suction. This it does in the first hour of its existence as well as if it had been taught by example and had practiced the art for years. It is instinct which teaches the newly born child to breathe, to cry when it is hungry or otherwise uncomfortable, and to clasp with its tiny hand the finger that is put into it.

It is instinct that teaches a bird how to make its nest after the way of its kind, to sit upon its eggs until they are hatched, and to feed the young with their appropriate food. This may seem to many of my readers a needless statement, but even in one of the learned societies of London I have heard a speaker assert that the power of building the nest was not an innate quality, but was communicated to the young by their long observation of the nest in which they were reared. That such an hypothesis is utterly absurd may be seen from the following facts.

In the first place, although the young pass their first few weeks inside the nest, they do not see the outside, neither can they possibly learn from their parents where the materials were obtained and the mode of putting them together. Each species, moreover, adheres to the habits of its kind, so that a chaffinch, if bred in a redstart's nest, would build the nest of a chaffinch and not that of a redstart. There have been countless generations of cuckoos, but, although every one of them was bred in the nest of a foster-parent not of its own species, not one of them has learned to build a nest for itself, but, when it becomes a mother, is taught by instinct to lay its eggs in the nest of some other bird.

Take the case of insects. Instinct teaches the silkworm to make its cocoon, to wait there until it is developed into a moth, and then to force its way into the world. It has never seen a cocoon before, so that it could not learn by imitation. Its mother died long before it was hatched, so that it could not learn by instruction. But, taught by instinct, it forms its cocoon exactly as did its parents whom it never saw, and as will its offspring whom it never will see.

All practical entomologists are familiar with

many instances of pure instinct on the part of insects. One of the most common is furnished by the well-known currant moth, or magpie moth, as it is sometimes called, which may be seen any summer day flitting about the currant bushes, seeking for a convenient spot in which to place its eggs. It is a very conspicuous insect from its mottled yellow, black, and white wings, and is remarkable for the fact that the perfect insect, the pupa, and the caterpillar, all possess the same colors.

The caterpillar belongs to the group which is scientifically termed Geometridæ, or earth-measurers, and popularly loopers, on account of the manner in which they walk, not crawling like other caterpillars, but drawing up their bodies in the middle into a staple-like shape, and so advancing by successive steps, stretching themselves straight and drawing themselves into a loop alternately.

All these caterpillars are provided with spinnerets and silk-producing apparatus, by means of which they can save themselves if they fall from a branch—an accident to which their way of walking makes them peculiarly liable. As they proceed, with the head and tail drawn closely together, they attach a thread to the object on which they are walking; and when they stretch forward the body to take a new hold with the front legs, they draw out a corresponding length of silken cord. If they should fall, they are brought up by the cord; and if danger should threaten, they let themselves down to the ground, and regain their position afterward by climbing up the suspended cord. Sometimes a knowing bird has been observed to take advantage of this habit, and to shake the branches until the caterpillars had lowered themselves to the ground, when he descended and ate them at his leisure, instead of hunting for them among the branches.

These caterpillars are hatched toward the end of summer, and feed for some three or four weeks, when they make preparations for the coming winter, which they must pass in a state of somnolescence. Let us watch one of them at this period of its life. Its home is within a leaf of the currant or gooseberry, the edges of the leaf being drawn together and fastened by silken cords. But, before doing this, the caterpillar ties the leaf to the branch by several strong silken bands attached to the stem.

This process completed, the caterpillar goes into its winter-quarters, and sleeps undisturbed until spring. In process of time, the laws of nature loosen the leaf from the branch: it can not, however, fall, being tied by the silken cords, and so it only hangs suspended, and swings about safely in the wind until the following spring.

Now here is a remarkable example of instinct pure and simple. It is utterly impossible that the caterpillar should know that the leaf would fall in the coming winter-time, and that the threads would keep it safely suspended until the warm weather of the following year.

Indeed, it is absolutely impossible that the creature should even know that there was such a season as winter, or that it would be obliged to live in the state of hibernation for some six months. When it again retires into quiescence during its pupal state it does not act in the same manner, but merely slings itself to the branch by its tail, previously spinning around it a slight cocoon by way of protection.

In both cases instinct, and instinct only, dictated its actions. In the one case it fastened the leaf to the bough, without knowing that the leaf would soon fall; in the other it slung itself to the branch, without knowing that during the warm days of summer it will need no protection from the elements and little from enemies.

It is instinct which teaches the newly hatched chicken to run about and peck up its food for itself, while instinct teaches the young pigeon to sit still in the nest and wait until fed by its mother. Ducks, though hatched under a hen, will instinctively make their way to the water; while chickens, though hatched under a duck, will instinctively keep out of it. Instinct throws a monkey into the most abject terror at the first sight of a serpent; while instinct teaches the secretary-bird, at first sight of a serpent, to kill and eat it. Instinct, and not parental instruction, teaches animals to select such food as suits them, and to reject that which would injure them. There are certainly some cases where instinct fails, as, for example, cattle who poison themselves by eating the leaves of the yew. But, in these instances, the cattle are domesticated, have not been obliged to depend wholly on their own efforts for procuring food, and their instincts have in consequence lost much of their power.

It is instinct which directs with unerring accuracy the cormorant to plunge into the water and to capture the swift fishes in their own element. It is instinct which tells the mole to find its food beneath the earth, and the swallow to catch the flies in the air. The swallow never tries to catch fish, nor the cormorant to chase flies, each being endowed from birth with the power of knowing its proper food and the means of obtaining it.

It is instinct which teaches the dragon-fly, an active inhabitant of the water, and the drone-fly, an absolutely inactive inhabitant of the mud, while in their larval states, to take to their wings as soon as they have attained their perfect condition, and to dart through the air quicker than the eye can follow them. They use their wings at once with as much skill as if they had learned under skillful teaching and with long practice.

It is instinct, and not reason, that forces the birds to migrate, and which guides them in their long journeys.

Man, as well as the lower animals, has his instincts; but, as he is able to bring most of them in subjection to his reason, very few of them are apparent. Some, however, remain and assert themselves throughout the whole of human life.

Reason differs from instinct in the widest possible manner, the former being an exercise of the will, and the latter independent of it. Instinct is implanted at birth, while reason is an aftergrowth of the mind. Instinct requires no exercise of thought, while reasoning may be briefly defined as a *deduction of a conclusion from premises.* This power is possessed by animals in common with ourselves, although not to the same extent; and it is by the superiority of our reason over that of the animals that we maintain our supremacy. Very often their deduction is insufficient, or their premises false; but the process is still one of pure reason, and has no connection with instinct.

With them, as well as with ourselves, reason often conquers instinct, especially in the case of those animals which are domesticated, and so develop their reasoning powers by contact with reason of a higher quality than their own. For example, if a hungry dog or cat be in a room where food is left unguarded, their instincts urge them to jump upon the table and satisfy their hunger; if properly trained, however, their reason restrains their instinct, and, no matter how hungry they may be, they will not touch the food until it is given to them.

I had scarcely written these words when I received the following anecdote, which shows the power of reason over instinct in exactly the manner which I have mentioned:

"A cat of ours once showed great self-denial. She was a terrible eater of small birds, chickens, etc., and therefore, when on one occasion she was found to have passed the night in our aviary of doves, great was the alarm. However, on inspection, not one dove was missing; and though she was asleep in an inner cage, close to a nest of young doves, she had not touched a feather. What made her conduct the more remarkable was the fact that on being released she ate ravenously."

It is just the same with ourselves. A child that has been well brought up can be left with perfect safety alone with any kind of dainties, the parents having taught its reason to conquer its instincts. Whereas a spoiled or ill-bred child, which has been suffered to allow its instincts to be paramount, will be sure to fall upon the coveted dainties as soon as it is left alone, and probably to make itself very ill. Surely the conduct of both the animal and the child is identical.

In the human idiot we have too frequent examples of the terrible power of instincts or propensities, as they are sometimes called, when the reason is insufficient to counterbalance them.

Almost any animal can be thus trained to subject its natural instincts to its reason. I have a letter from a lady, who writes that she has a pig which for good manners and cleanliness is as fit for a drawing-room companion as any lap-dog.

The distinction between reason and instinct is strongly defined in the conduct of a dog who possessed both qualities in a very superior degree.

The animal in question was named "Don," and was in his master's opinion the "prince of pointers." His scent was extraordinarily keen. For example, one day, when out shooting, he suddenly came to a point, and stood like a rock. His master went up to him; but no game rose, and still the animal continued to point. His master walked on in the indicated direction, until he was stopped by a stone wall, and on looking over it he saw a hare closely crouched to the ground. The keen scent of the dog had detected it in spite of the intervening wall.

As often happens, birds got wild toward the end of the season, and used to rise while out of shot. Now "Don" knew the range of the game as well as his master, and invented a singularly ingenious mode of literally circumventing the birds.

His peculiarly keen scent enabled him to detect them at a considerable distance, so that they would not be afraid of him. Instead of going directly toward them, "Don" used to circle around them, gradually contracting his lines until he came within range. He would then look back at his master, as if to say, "It's all right, we have them now;" and so they had.

Here we see both characteristics developed to the fullest degree, the lower being controlled by the higher, and used as its tool. The singularly keen scent was purely instinctive, and had nothing to do with reason. But the reasoning powers of the animal enabled him to employ his instincts in the service of his master. First, he had observed that the gun was never used beyond a certain range, and had come to the conclusion that beyond that range birds could not be shot.

Then he had observed that when birds were wild they rose out of distance, and so set himself to invent some plan by which they would not take alarm while out of shot. The device which he practiced was exactly that which is at the present day employed by the hunters of South America. If they see a partridge in the plain, they ride around and around it in ever-narrowing circles. The bird lies closely crouched to the ground in hopes that it is not observed, and the horseman at last approaches so closely that he is able to kill it with a blow from the metal handle of his whip.

Fishes are not supposed to be possessed of much reason; yet every angler knows that all the powers of his mind are taxed before he can induce an old and wary trout to take his bait, or, when he has succeeded in hooking the fish, to prevent it from breaking his line.

The natural instinct of a fish teaches it to fly from man, and we all know that even our shadows on the water will frighten away the fish and destroy the angler's hopes of success. Yet I know a pond full of gold-fish which are quite tame, and which, when they see a human being at the side of the pond, come toward him instead of being alarmed. If a little rippling be made on the surface of the water, they come crowding to the spot, that being the signal for food; and so perfectly confiding are they that they will take bread or biscuit out of the hand, and if the hand be kept under the water, one or two of the fishes will presently be nibbling at each finger.

Here then is an example of the instinct, which urges them to flee from man, being overcome by the reason, which tells them to approach him. I have seen an electric eel fed in just the same manner. The creature was blind; but it at once recognized the ripple, coiled itself around the spot where the water was agitated, and with a shock killed a fish which the keeper had placed there. At the British Museum there are now some Axolotls kept alive in a glass vessel. They are sluggish creatures, mostly lying at the bottom of the vessel; but if the water be agitated, up they come with open mouths, expecting the strip of meat with which they are fed.

This conduct is a distinct deduction of a conclusion from premises, and, so far from being dictated by instinct, is absolutely opposed to it. If the reader will keep in mind the definition of reason, he will see that, in all the anecdotes which are narrated in this and the two succeeding chapters, reason, and not instinct, is the motive power.

The following account of a tame "Horned Toad," or "Horned Frog," as the animal is called, is written by one of my brothers. I may first state that the creature in question is neither a toad nor a frog, but a lizard belonging to the great family of the Iguanas. Its native popular name is Tapayaxin, and it is known to science as *Phrynosoma Blainvillii*. The former of these names is composed of two Greek words signifying toad-bodied, and is given to the creature on account of its flat, toad-like aspect.

"The Horned Toad, so called by the people of the regions inhabited by this curious reptile, is a very oddly shaped lizard, measuring when full grown about six inches in length, of which the tail occupies one and a half inches, and three inches across the back, which is enormously wide and flat when compared with the little and elegant forms of the lizards in general.

"The head, back, and tail are thickly planted with spines, which in the full-grown animal look exactly like those of the black-thorn. The head from behind the eyes radiates spines; the back is covered with them, some large and some small. The two edges of the belly are set like the teeth of a saw, as is also the tail, which appendage is short for the size of the animal, and tapers from three quarters of an inch at the base to a point at the extremity, being a distance of only an inch and a half.

"This lizard, probably from its form, is not nearly so active as its race generally are—even when disturbed seldom running more than three or four feet, and then stopping close to some stone or root, to which instinct teaches it it bears a close resemblance, and trusts to that resemblance to escape detection, in which it often succeeds, as in such cases none but an eye educated in observation can trace the fugitive, or detect in the apparent root or stone a living reptile; on these occasions a quick grasp of the hand will mostly secure it alive.

"The facility with which these strange creatures are tamed is almost ludicrous in its effects. When seized in the hand, it endeavors to escape by repeatedly pressing its head against the detaining fingers of its captor, in the hope that the spikes with which it is armed will effect its deliverance; but then if the head is allowed to protrude from the hand and gently stroked, and the under-jaw treated in like manner, in less than a minute the eyes close and the creature is asleep; and it will be found, upon awakening, that the timid, shy lizard is completely tamed. This curious process I have tried on some eight or nine specimens without a single failure.

"When thus tamed, these lizards make the most engaging pets possible, their forms are so strange, and their actions so quaint and old-fashioned in the extreme. They are very chilly creatures, reveling in the mid-day sun, and hiding away in some warm corner when the sun goes down; in the wild state they scrape a small hole in the sand, heated by an almost tropical sun, and lie there during the night, until the warm rays of the morning sun again arouse them into activity.

"This habit showed itself to me in a very curious manner. I had caught some seven or eight specimens, and put them in a box with about an inch of sand at the bottom, where they ran about merrily enough during the day; but next morning, when I looked to see how they were getting on, not one was visible, and I naturally supposed that they had taken French leave and escaped. But after the lapse of an hour or so I heard a scratching in the box; and on looking in, there were all my little friends, some running about, others still half buried in the sand. This led me to suspect their habits, and so next morning, just after sunrise, I went to the place most frequented by them, and sat down patiently to watch for them. In about half an hour my eye caught a movement in the sand about half-a-dozen yards to my right, and, after a moment or two, out came a lizard, and before an hour had passed I had seen four come out of their sandy beds.

"I hear from those who have traveled on the greater part of this continent that this lizard is only found in the Sage-brush district, and *never* near water or damp places. It is also stated that one of these reptiles may be placed in a bottle, corked and sealed up for years, and will be as lively at the end as when first put in. I have commenced an experiment on this subject. On the 1st of September I placed four specimens in different bottles, corked, sealed, and then, over all, several layers of tinfoil tightly pressed down.

"I have had one very large specimen living at large in my bedroom for the past six weeks; during this time he has on several occasions gone out of doors on fly-catching expeditions, but always returned to sleep under an old piece of cloth in one corner of the room; and even when outside, where a run of a couple of yards would give him liberty, he will always allow me to pick him up without trying to escape.

"But of all the amusing proceedings on his part is his way of catching flies in the room.

"During the latter part of the day the sun shines through a hole in the shutter of my bedroom, and makes a nice warm spot on the floor alongside one of my portmanteaus, and on this spot the flies 'most do congregate;' so my little pet, who is not quick enough to catch the flies in fair chase, climbs on the top of the portmanteau, and, lying half on and half off, watches his opportunity, and woe to the unfortunate fly that settles below him; the instant the fly is quiet the lizard gives a few preliminary curls to the tip of his tail, just as a cat does when watching a mouse, and then tumbles down bodily upon the heedless fly, cuddles his prey between his fore legs and chest, and then, bending down his head as far as possible, allows the fly to struggle out of his embrace, when with one quick motion of his tongue the poor fly has disappeared. After a moment's rest, up he climbs again, and is ready to repeat the process.

"I have once counted seven flies caught in this manner within an hour, and, during the whole of the time that I have had him, I have only seen him miss twice in catching the prey upon which he had fixed his eye.

"As the nights are getting colder, I notice my pet is daily becoming more lethargic in the morning, and from this assume, in all probability, this species hibernates during the winter. I hope to be able to make some few experiments upon this subject during the coming winter, and the results, if any, shall be duly forwarded for publication."

The writer of this notice sent me a Tapayaxin *by post*. It arrived in perfect health, and lived for some time; but I have no hot-house, and a severe winter killed it.

The reader will probably have observed that in this mode of catching prey the creature was guided by reason rather than by instinct. It had observed that the flies were in the habit of settling on the spot which had been warmed by the sunbeam, and so took advantage of the portmanteau as a post of vantage whence he would leap, or rather fall, upon his prey.

If there be any animal in which we should look for reasoning powers, it is the dog. I propose, therefore, to give a few original anecdotes of this animal, in all of which the power of reason will be evident. In the course of this work many other anecdotes will be related, both of dogs and other animals, in which their power of reason will be shown; but as the anecdotes have a more distinct bearing upon other attributes, such as love, hate, sympathy, generosity, etc., they will be placed under those respective heads. In the two anecdotes which follow it seems as if the man and the beast had almost changed places. At all events, even if the animals did not possess more reasoning powers than the men, they exercised those powers which they did possess to a better purpose.

"I had a friend who possesses a little black-and-tan English terrier. His master had the misfortune not only to prefer two glasses of grog to one, but greatly to prefer three or four, with the usual consequences. On one of these occasions he beat his dog severely, and from that time the dog, whenever there was a recurrence of the *fourth* tumbler, went and hid himself in the cupboard, never showing himself until the effects had passed off, and his master was restored to sobriety.

"I know of another dog, a Scotch terrier, whose master is extremely fond of him, and the attachment is mutual. At times this gentleman exceeds the bounds of prudence, and, when he does so, the bout lasts for two or three days. Although on these occasions he is quite inclined to fondle and make much of his dog, the animal will not go near his master, nor even look at him, but shuns him in every way, and keeps aloof until his master is restored to a perfect state of sobriety."

There was a Scotch terrier dog who lately died, to the very great sorrow of his master, an officer in the 45th Regiment, and the very great rejoicing of his master's friends. He was good enough to honor me by admitting me among his friends—the only person not belonging to the family to whom he extended that privilege. His name was "Mess," which was a military abbreviation of "Mesty," which was an abbreviation of Mephistopheles, the name being given to him in consequence of his temper, which really deserved the name of infernal. No one, except his master, his master's family, and an exceptionally favored servant or two, could put a hand on him without being bitten. I know a learned barrister who has been kept in bed until a very late hour in the morning because "Mess" had come into his room when the servant brought the hot water and would not allow him to get up. As long as he lay still in bed, "Mess" sat quietly on the floor; but at the least movement "Mess" sprang up with a menacing growl, flashing eyes, and gleaming teeth, and the unfortunate guest had to subside again, unable even to ring the bell for help, and anxious that his host and hostess must be waiting breakfast for him and chafing at his laziness.

One day I paid a visit to "Mess's" master, not knowing any thing about the dog, and not seeing the dog when I arrived. Being accustomed to an early walk before breakfast, I started off as usual on the following morning, and on returning met a little procession, consisting of a nurse-maid leading a donkey, on which were the two daughters of my host in panniers, and a remarkably fine Scotch terrier, which was trotting along in front. As soon as he saw me, the dog sprang forward, and I, not knowing any thing of his character, and thinking that he wanted a game, stooped down, patted him, rolled him on his back, pretended to box his ears, put my hand into his mouth, and, in short, let him have his game. The nurse-maid stood by almost paralyzed with horror: but why she should be frightened seemed rather mysterious.

On coming to breakfast I spoke in high terms of the splendid dog with whom I had enjoyed a game, and the host was almost as horrified as the nurse had been. Not until then did I hear about the dog's temper; but, whatever it was, it was never displayed toward me, and I believe that I am the only person not belonging to the family who was ever allowed to put a hand on him. I may mention that a life-sized portrait of "Mess" was taken in crayons by Mr. Waterhouse Hawkins, and occupies a place of honor in his master's dining-room.

Some years ago "Mess" and his master were stationed at Parkhurst, where was a dépôt. Although several regiments were represented, "Mess" perfectly knew the green facings of his own regiment, and would recognize men belonging to it, but no others. This, by the way, was the more curious, as all the troops wore the scarlet coat. He had a way of being present at the morning parade, and then going off to the barrack-rooms to breakfast. He had arranged in his own mind a regular series of rooms to be visited; and if the men succeeded in decoying him into a room which did not correspond with the day, he bit somebody and went off to the right room.

There are many officers and men of the 45th who perfectly recollect "Mess" even after the lapse of several years.

Once, while home on leave, his master was taken with a fit of illness, "Mess," as a matter of course, keeping guard. In the course of the night the necessary medicine was brought by the patient's mother, who wore a rather elegant nightcap, reserved, as she used to say, in case of fire. The medicine happened to be peculiarly distasteful, and the patient gave an involuntary shudder. Whereupon "Mess," thinking that his master was being injured, flew at the lady, and never afterward would endure the sight of a feminine nightcap.

"Mess" was good enough to extend his friendship to his master's father, a surgeon, and condescended to accompany him on his rounds, sitting in great state on the box. One day he fell off as the carriage started, and the wheels went over him, breaking one of his legs. He would not allow himself to be touched, except by the surgeon's hands; and to him he was quiet and amenable, allowing his leg to be set and laid in splints without showing the least anger, and being evidently grateful for the services rendered to him. The leg rapidly recovered, and "Mess" was at his master's country-house when the surgeon came to pay his son a visit. No sooner did "Mess" see him than, although his injury had long been healed, he began to limp, went to his old friend, rolled over on his back, and held up his leg. Nor would he desist until a handkerchief had been tied around the leg and some water poured over it. Afterward, when he happened to injure a paw, he went of his own accord to the surgeon, held up the damaged limb, and asked for help as plainly as if he possessed human language.

We shall hear more of "Mess" in succeeding pages, but meanwhile it is impossible not to see that the actions of the dog proceeded from real reason. Sometimes his premises were false, as in the case where he kept the guest in bed, or when he must needs have the sound limb dressed; but there is no doubt that he did draw a conclusion from premises, and that therefore he possessed reason.

A lady of my acquaintance once saw a curious instance of reasoning in a toad.

She was sitting in a garden, when she saw something alive moving along the base of the wall, which was an old one and full of crevices. The object proved to be a large toad, which was examining the wall in a most systematic fashion. She saw the creature raise himself on his hind legs, peer into a crevice first with one eye and then with the other. Then he tapped the wall with his paw, and pushed it into the aperture. Evidently dissatisfied, he went away, and tried another crevice in the same manner and with the same result. A third, however, was larger than the others; and this seemed to be to his taste, for he slowly drew himself up the wall and disappeared into the crevice.

It was evident that the creature knew his own dimensions, and was taking measurements of the crevices in order to find one that would allow him to enter. Toads, by the way, possess sufficient reason to be easily tamed, and to come at a call. My children generally have some tame toads in the summer-time, and are in the habit of carrying them around the garden, and holding them up to let them catch the flies and other insects that settle on the flowers. The creatures are so accustomed to this mode of being fed that they do not require to be held, but sit quietly on the open hand.

It is very curious to note how the reason of the lower animals suddenly fails just where least expected. My bull-dog, "Apollo," an animal of peculiar intellectual powers, once displayed a singular example of this sudden failure.

I was walking out, with Apollo as usual at my heels, when I met a party of friends, who began to ridicule the dog, saying that he was of no use except at a dog-fight, and could not even fetch or carry. I answered by throwing my stick, a heavy "Penang lawyer," over a high park fence standing on the top of a steep bank. Apollo dashed after it, and, being lithe and active as a greyhound, he sprang up the bank and fairly leaped the fence, just helping himself over with his legs.

Presently we saw his round head come up on the other side of the fence, the stick being in his jaws. It was so heavy that he could not even get his fore legs on the fence, and so he ran along the inside trying to find an outlet. As the fence had been recently repaired, he could not find an exit, and straightway set about making one. He put down the stick, and deliberately bit a hole through the fence, tearing away the oak planks as if they were pasteboard, until he had made a hole through which he could pass. He went through the hole, put his head into the field, took the stick in his mouth, and tried to pull it after him. As, however, he had grasped it by the middle, the stick naturally resisted his efforts.

I thought that the dog would be sure to take the stick by one of its ends, and so pull it through; but, instead of doing so, he went back into the field, and tore away the fence till he had made a hole large enough for the stick when held by the middle.

This story is the more remarkable because other dogs, certainly not of greater mental calibre than Apollo, have resorted to that very simple mode of getting out of a difficulty.

For example, I have a letter before me in which is an account of a dog who had been sent into the water after a wooden rail about eight feet long and several inches wide. The animal took it by the middle, and swam with it to the only place where a landing was practicable; but, finding that there was not sufficient space for the long rail, he swam out again, turned the rail around, took one end in his mouth, and so brought it ashore. Presently his master threw the rail into the water again, and this time the dog took it by the end at once in order to bring it in, never seizing it by the middle after his first failure.

CHAPTER III.

REASON—(*continued*).

History of a Friend's Dogs.—" Pincher " and his Mistress.—" Pepper " and the Velvet Cat.—The Maltese Dog and Lady C.'s Carriage.—" Joey " and the Alarmed Household.—Joey's Last Days.—Dogs Discovering Lighted Gas and Unfastened Doors at Night.—The Cat Detective.—Dogs Understanding the Use of Money. —The Penny and the Red-hot Iron.—The Margate Dog and the Baker.—The Suicide's Dog.—The Hat and the Walk.—Hide and Seek.—A Too-conscientious Dog.—The Terrier and her Hiding-place.—" Bosco's " Curiosity Gratified.—A Gallant Rescue and Deserved Castigation.—Acquisitiveness in a Dog.—Lord M.'s Dog and the Runaway Horse.—The Retriever and the Hedgehog.—Courtesy to Ladies.—An Ass too Clever to be Kept.—Various Modes of Opening Doors.—The Clever Mule.—A Morning Caller.—The Monkey, the Cage, and the Strap.

A LADY who has a great fondness for animals, especially dogs, has kindly sent me a few anecdotes relating to traits of character among her pets. The following have been selected as examples of reason in the dog, though other traits are also manifest.

" Poor old Pincher! His name was most old-fashioned and unaristocratic, and he was one of the occupants of our stable-yard, and never allowed to enter the house. His education was totally neglected and uncared for. He was a middle-sized, smooth-haired, black terrier, and had acquired some peculiar ways of his own.

" In his time we were in the habit of spending about two days per week at our country-house, ten miles distant from Canterbury. Pincher generally accompanied our carriage, and seemed to enjoy these country days as much as any of us. On one occasion, home-engagements had prevented us from paying our accustomed visit to Harnden. Pincher disapproved of the alteration, as he started off with the carriage as usual; but when he found our destination was not Harnden, he refused to follow, but turned off to the house, went the whole distance (mystifying the servants there, who expected us to follow), remained there until evening, and then returned home.

" Years later the poor dog became too old to accomplish the whole distance. He could not walk, and would not ride; so he adopted the expedient of going about half-way with us, always waiting at the same spot until our return, and then following us home."

" A little Scotch terrier, named ' Pepper,' one of our former pets, was, like most of his relatives, a capital fellow for hunting a rat, a cat, or a mouse. He was our companion when calling on an old lady, where I thought we could take him without any fear of his hunting propensity causing annoyance, as I knew she had no living pet of any description. We had scarcely entered the spacious drawing-room, when, from underneath an Indian cabinet at the extreme end of the room, our dog Pepper saw two large, glassy, yellow eyes glaring at him with more than natural ferocity. Without waiting to use his power of scent, he rushed fiercely on his imagined foe, which fell lifeless at his feet, Pepper retreating to our side, hanging down his tail, and looking more like the vanquished than the victor.

" Do any of my readers remember those now unmade cats of pasteboard and black velvet texture, those now non-existent ornaments of former days? Such was Pepper's foe. Dogs know well enough when they are the objects of ridicule, and, finding we were all laughing at his discomfiture, he returned to the velvet pussy, and in playful mood carried her around the room, evidently wishing to hide his mistake by convincing us that it had only been a sham fight from the beginning."

The action of the dog here is very human, and it behaved just as a clever child might be expected to do when it had been deceived, and was afraid of ridicule. In the next anecdote the reasoning powers of a dog are seen to be at fault, as they sometimes are with human beings.

" The dog which we now have, though not an equestrian like his predecessor, is exceedingly fond of carriage drives; and if a well-appointed carriage should draw up, he will often stop, and look up most pleadingly at the coachman to have the door opened. Of course he has had many drives with us in cabs, but he never of his own accord enters one of these vehicles.

"We have been lately staying in town, and the day after our arrival we went out, followed by our little dog. We had just passed through one of the large squares, when we missed our favorite. With only a faint hope of finding him, we retraced our steps to the square, where a handsome carriage, with coachman and footman, was drawn up at the door of one of its most stately mansions. We asked the footman if he had seen a little white Maltese dog in the square. 'Yes, madam: as soon as Lady C—— got out of the carriage, he jumped in so quickly that I had not time to prevent him, seated himself on the cushion, and defied me to remove him.'

"There he was, evidently waiting for us. Just at that moment her ladyship came out of the house, accompanied by her little pug-dog. In jumped the pug, down jumped the Maltese, and there was a sharp fight, which was ended by my removal of the usurper. We were only too glad to find him again, and Lady C—— said, good-naturedly, that he deserved a drive for his determination."

"A few years ago we left our household, in the old city of Canterbury, in the charge of a man and his wife, who also undertook the care of a little Maltese spaniel, named 'Joey.'

"On one rough, blustering December night, when the inhabitants were in their profoundest slumbers, a tremendous 'bang' resounded through the house, awaking all its inmates, including Joey, just as the cathedral clock struck the midnight hour. Men and maids rushed hither and thither; but no evident cause could be found for the alarm, every door and window being perfectly secure.

"After the first disturbance had subsided, Joey returned to the quietude of his own basket, with evident disgust at the unusual and, in his opinion, uncalled-for commotion, and refused to take any part in the search.

"At last he was forcibly put into the garden as an advanced guard, but he flatly declined to move a step. All joined in upbraiding him. 'Joey was lazy!' 'Joey was a coward!' 'Joey was no use as a watch-dog!' And one of the servants expressed an opinion that he had been drugged by thieves, and that she had noticed a drowsiness on the preceding day. Next morning it was found that the large old-fashioned clock in the lobby did not strike the hour as usual. The fact was, the weight which drove the striking part of the machinery had broken away from its cord, had fallen into the bottom of the case, and had produced the sound which had startled the house."

The dog evidently knew that no danger was signified by the sound, and so declined to trouble himself about the matter. This was the more remarkable, as he was the wariest and most suspicious of dogs. He would never compose himself to sleep unless the shutters of the garden door were properly closed, and used to bark and growl at the door until it was made secure.

"In a former letter to you I mentioned my little dog Joey.

"The last summer of his life we left him as usual in the care of a man and his wife; but this time, unfortunately for the dog, their son George, a boy of fifteen, was at home for his school holidays. On our return, after an absence of some months, no little Joey welcomed us, and no one can tell how we missed his merry voice. Was Joey dead or ill? No; he was only shut up in a room up-stairs until after our arrival.

"We thought this to be rather a mysterious proceeding, and, on our entering the room, the poor little animal rushed to meet us, and then fell down powerless in a fit. The first time that George came into the room, Joey walked up to him, stared him in the face, and commenced a series of growls, looking at us every now and then as if to ask whether we understood him.

"Whenever the boy entered the room this scene was repeated, and, even if we took Joey in our arms, he continued to growl, and seemed as if he thought that we ought to growl also. At the time we could not interpret his meaning; but we afterward discovered that the boy had given him a blow on the head, which caused him to have fits whenever excited, and at last caused his death. How plainly did he tell us who had injured him!"

In neither of these cases was instinct in the least concerned, the whole proceedings being dictated by reason, and reason alone. In the first instance the dog knew whence proceeded the sound which had alarmed the house, reasoned with himself that there was no cause for alarm, and, though he would have been in a paroxysm of barks if danger had really impended, he went back to his couch, and declined to trouble himself. In the second case the poor little creature, not possessing human language, tried to make his friends understand, by a language of signs, that he had been injured by the boy. The language was singularly expressive, and would have been at once understood, were it not that his mistress was herself so kind to animals that she never suspected that any one could be capable of doing the dog a willful injury.

I know several instances where domestic animals have discovered that there was something wrong in the arrangements of the house, and have called attention to it. There is a little dog belonging to one of my friends, who one night became very importunate, pulling the skirt of his mistress's dress, and insisting on her returning down-stairs. She was rather alarmed; but the dog drew her to the greenhouse door, which he evidently meant to be opened. On unlocking the door, she found that she had forgotten to turn off the gas. The little dog had been accustomed to see the gas turned off before the family went to bed, and was too conservative to allow any change.

Here is a similar example, which was communicated to me by a lady. "Did I tell you that my dog Tiny once found that the housemaid had forgotten to shut a closet door in a bedroom at the top of the house? He came to me, made me follow him, and showed me the open door."

Cats are not generally considered as house-guardians, but that they can act as such the following anecdote will show.

A lady had a very strong objection to "followers," and forbade her servants to receive a man into the house. One evening she was sitting in the drawing-room, when she heard the cat mewing and scratching at the door, as if for admittance. She opened the door; but the cat would not enter, and evidently wished to be followed down-stairs. She then descended the stairs, and led her mistress into the kitchen, where was the obnoxious "follower."

This anecdote shows also that the animal must have been able to understand human language, or otherwise she could not have known that her mistress had forbidden strange men to enter the house.

In the two following anecdotes the action of the dog can only be attributed to reason, and that of no mean character.

The first anecdote was sent to me by one of the principals in a well-known engineering firm.

"I once lost a sovereign in a bet which I made that a wonderful little dog would not take a penny off a red-hot bar of iron. The dog belonged to an ironmonger at Knighton, Radnorshire. The dog was in the habit of searching for pence purposely hidden in the shop, and, when found, taking them to a baker's shop and getting buns in exchange. He quite knew the right-sized bun, and used to keep his paw on the penny until he got it.

"A bar was heated red-hot, and no sooner was the penny laid on it than the dog, without the least hesitation, dashed at it. By some means which I could not see, because it was done so quickly, the dog knocked the penny off the bar, and then sat down quietly by it until the coin was cool. His look of perfect self-satisfaction was most absurd."

Some years ago there was a dog at Margate which also knew the use of money. He used to beg for pence, and take them to a baker to be exchanged for biscuits, at a shop in the narrow, hilly lane which is pleased to assume the title of High Street. One day the baker, wishing to see how the dog would behave if he played the animal a practical joke, took his penny and gave him a burned biscuit. The next time the dog had a penny he took it to the baker as usual, showed it to him, and then went off to another baker who lived nearly opposite. This he afterward did invariably, showing the penny to the baker who had offended him, and then transferring his custom to the rival on the opposite side of the narrow street.

The whole of these proceedings were dictated by pure reason, and instinct had nothing to do with them. It was, in fact, doing on a small scale precisely what the dog's master would have done on a large scale if a tradesman had taken his money and given him a bad article for it. He would have withdrawn his custom from the offender, and given it to another man who he thought would serve him more honestly.

No one can say that instinct had any thing to do with these proceedings, the dog in each case deducing a conclusion from premises, and deducing them rightly. Had a child acted in the same manner, we should have thought it a very clever child; but we certainly should have attributed its action to reason, and not to instinct; and I do not see that we have any right to attribute reason to the one and to deny it to the other.

We are familiar with many instances where dogs have tried to assist their fellow-creatures, whether human or belonging to their own kind. The following history of a suicide's dog was sent to me by a Scotch lady, who takes a great interest in dogs.

"There is a cottage called 'Blaw-weary' on the farm of C——, the property of the Marquis of T——. This cottage is just on the other side of our march-fence, about half a mile from our house on the west.

"A shepherd lived in Blaw-weary some four years ago (about 1868), and one day he and his collie dog went out early in the morning, according to their custom. At breakfast-time the dog returned alone, looking miserable, and would eat nothing. After remaining a few minutes, he went out again; the man's wife, who was at home, suspecting nothing. At dinner-time the dog came back again, also alone, and 'banged through the hoose,' as his mistress said.

"Presently he went out, and soon came in again, making piteous efforts to attract attention. The assistant shepherd followed the dog, and was taken straight to a small clump of trees in the neighborhood, on one of which was hanging his master, quite dead. The poor dog would not allow any one to touch the body; and it was not until after he had been overpowered and led away that the corpse could be removed from the branch on which the wretched man had hanged himself."

Here we have a story which is exceedingly valuable, as it shows not only that the dog possessed reason, but is another proof that the reason will sometimes suddenly fail exactly when it seemed to have been most successful. The poor animal had evidently witnessed his master's dying struggles, and, feeling himself unable to help, had gone to his house for assistance. Having obtained that help, however, he could not understand that any one could touch his beloved master without intending to injure him. Many medical men have met with similar experiences, the dog bringing assistance to his helpless master, and then not suffering any one to touch him.

Probably the animal felt that his master was dead, and that no one could restore him to life.

The following stories illustrating the reasoning powers of dogs have been sent to me from Scotland, where dog-nature seems to be better appreciated than in England.

"A retriever, named 'Bevis,' an old favorite of our own, was in the habit of going for a walk before breakfast with my father. One morning it so happened that my father did not intend to take his usual walk. Bevis soon became very impatient, and, seeing no signs of his master, he got upon a chair in the hall, took his master's hat off its peg, carried it up to his room, and then scratched at the door for admission. As soon as the door was opened, in walked Bevis, laid the hat at his master's feet, and pushed his nose into his hand. It was entirely his own idea, as he had not even been taught to fetch a hat."

"Another dog of ours, a little Maltese poodle, named 'Pop,' was unusually full of tricks and oddities.

"He was fond of a game at hide and seek, a key being hidden for him, while he buried his face in the sofa-cushions. Sometimes he would be guilty of cheating, and would slyly peep out to see where the key was being hidden; but when reproached with the two simple words, 'Oh, Pop!' he would put down his head again, and be very much ashamed of himself."

The reader can compare with this story several anecdotes of a similar character related in the chapter headed "Humor."

The following anecdote, which was sent by the same correspondent, affords a good example of wrong reasoning, i. e., drawing an incorrect conclusion from the premises.

"A collie dog, named 'Moss,' belonging to a farmer, had excited the admiration of a drover who was helping the shepherd to bring home cattle to the farm. The drover asked to be allowed to borrow Moss for a few days, to help him in getting some cattle from another market to Burntisland.

"The dog, being on friendly terms with the drover, went willingly, and gave his help in bringing the cattle on their journey. On their return, they had to pass the spot where the road to Burntisland branches off from that which led to his own farm; Moss refused to go any farther on the Burntisland road. Not only this, but he would not allow the drover to take the cattle any farther, and the man was at last obliged to let the dog deliver the cattle at his master's farm."

The process of reasoning is quite evident here. The dog had always known that any property of which he had been placed in charge belonged to his master, and consequently brought the cattle to his master's farm. His reasoning was correct enough, but one of his premises was false.

Here is another example of reasoning in a dog. Several successive litters of puppies had been taken from their mother, a little terrier. When the next litter was expected, she left the house, and was not seen again for some time.

At last she returned, bringing with her in great pomp a whole retinue of fine healthy puppies. It appeared that she had hidden herself in a rabbit-burrow, evidently knowing that, if she could only conceal her puppies until they were able to shift for themselves, no harm would come to them. The result proved that she had been perfectly correct in her interpretation of her master's character.

"One day my dog 'Bosco' wished to ascertain if the roast beef still stood upon the passage-table at the dining-room door. He stood on his hind legs, jumped up, but all in vain. So, after thinking a little, he ran a short way up-stairs, pushed his head through the banisters, looked down, and, after ascertaining that there was only pudding on the table, returned quietly to the parlor."

I have known a King Charles spaniel to act in very much the same way, except that in the latter case the dish which the dog wanted to inspect was on the dining-room table. After trying in vain to see what was on the table, he went out of the room, went half-way up the stairs, and so took a survey of the table through the open doorway.

The following anecdote was sent to me by a gentleman resident in the neighborhood of the locality where the adventure occurred.

"There is a water-mill, called Maxwellheugh, on the side of the road between Kelso and Teviot bridges. It is driven by a conduit of water from the Teviot immediately before its junction with the Tweed, and consists of two flats. The upper flat is on a level with the public road, and is called the 'Upper Mill,' while entrance to the lower flat, or 'Under Mill,' was reached by a cart-road descending from the highway.

"The first thing the miller did in the morning was to unchain the dog. He immediately placed himself on guard across the upper doorway while the miller proceeded with his work in the Under Mill. As soon as the miller had finished his work there, and removed to the Upper Mill, the dog, without being told, set off to the miller's house, and in two journeys brought his master's breakfast—namely, milk in a pitcher and porridge in a 'bicker,' tied up in a towel.

"On one occasion, when the Teviot and the Tweed were in flood, a little dog ventured incautiously into the Tweed, and was rapidly carried down the stream, struggling and yelping as it was hurried along.

"It so happened that the miller's dog, while carrying his master's breakfast to him, saw the little dog in distress. He immediately put down his burden, turned, and set off at full gallop down the stream. When he had got well below the drowning dog, he sprang into the river, swam across, and so exactly had he calculated the rapidity of the river and his own speed, that he intercepted the little dog as it was being helplessly swept down the current, and brought it safely to land.

"When he got his burden safely on shore, the dog, instead of displaying the least affection for it, cuffed it first with one paw and then with the other, and returned to the spot where he had deposited his master's breakfast, and carried it to him as usual."

How is it possible to refer the proceedings of this animal to mere instinct? Had a negro slave performed them, we should have used them (and with perfect justice) as arguments that so intellectual and trustworthy a man ought not to be the property of an irresponsible master.

The whole behavior of the dog is exactly like that of a burly, kindly, and rugged bargee, possessed of cool judgment and rapid action, willing to risk his life for another, and then to make light of the whole business. I was for some years in charge of a water-side parish, and knew many a bargee who would have acted exactly in the same way if a child had fallen into the river. He would have got the child out at the risk of his own life, and then, instead of waiting for thanks, would have boxed its ears soundly, rated it for interrupting him in his work, and then have proceeded with his journey as if nothing had happened.

The man would have been held worthy of the medal of the Royal Humane Society, and would probably have received it. The dog can receive no reward in this world: shall we say that he will receive none in the next?

The process of reasoning that took place in the dog's mind is as evident as if the brain had been that of a man and not of a dog. The animal exhibited self-denial, presence of mind, and forethought. Had he jumped into the water at once, he could not have caught the little dog; but by galloping down the stream, getting ahead of the drowning animal, and then stemming the current until it was swept within his reach, he made sure of his object; and no man could have acted better if he had tried to save a drowning child.

The following curious instance of reason joined with accumulation has been recently sent to me. I know the dog, and an odd, eccentric little being he is.

"Property of every description requires a certain amount of supervision, whether 'real or personal,' as the lawyers say, and has its attendant anxieties as well as its attendant pleasures; but I never saw any animal so impressed with the responsibility as our present little dog appears to be.

"Having been in our possession all his little life-time, the items of his personal property have gradually increased. At first he occupied the basket of his predecessor, which was taken up-stairs for him at night. After some time, another basket was purchased for the drawing-room, the old one retaining its place up-stairs. New things are always favorites with children, and this at first was supposed to be the case with our little animal: he would not occupy the old basket at night, so the new one was brought up at night and placed beside it. This was continued for a short time: when the old one was taken down, the new one only remaining up-stairs. This was not the right thing to do: he then refused to occupy the new one.

"I must confess to humoring his little peculiarities, so I fetched the old basket up, leaving *both* in the room. This was quite what he wanted, and gave evident satisfaction: he jumped into one, which he arranged comfortably, then performed the same operation in the other, and finally occupied both baskets at intervals during the night. He will now never compose himself at night until both baskets are in the room. One night I purposely removed his dish of water; he missed it, sat up begging on the spot it always occupied, and great was his delight on its restoration, although he had no wish to drink. I have given him duplicate property, and placed his baskets, water-dishes, etc., at different parts of the room; he never fails to go the round and inspect his property before fixing himself for the night, and most amusing it is to witness his anxiety until he has the whole of his goods under his own protection."

Here is an anecdote of important help rendered in a most unexpected manner. It was sent to me by the wife of the dog's owner.

"The late Lord M. had a very fine large black Newfoundland dog, called 'Neptune,' which used to be kept chained up in a court-yard outside the castle. Now Neptune was very fond of mutton and pork, and used to worry the sheep and pigs whenever he had a chance. He was consequently very seldom let loose or taken out, unless they were going to ride in some out-of-the-way district. On one occasion, in the autumn of 1856, Lord M. and Mr. H. were riding across country, accompanied by Neptune, when coming to a high bank with a broad ditch on either side, Lord M.'s horse refused to take it; so Lord M. dismounted, and, getting onto the bank, tried to lead him over it; but while so standing on the bank a gust of wind blew his hat off, and in trying to save it the bridle slipped from his hand, and the horse became loose.

"As quick as lightning, Neptune, who had apparently been most interested in the endeavor to get the horse over, sprang after the hat, and, catching it, jumped with it onto the bank, dropped it at his master's feet, and dashed after the horse, which was trotting off; and, before Mr. H. could overtake it, he had seized the bridle with his teeth, and held on, checking it till Mr. H. came up and took the bridle from him, when he appeared to express his pleasure by little short barks and a variety of gambols. What makes this a remarkable circumstance is that Neptune had never been broken in to fetch and carry, and had never been used as a retriever, or was known or seen to do any thing of the sort before that occasion."

In the following example of the conduct of a dog, it is impossible to see that instinct had any thing to do with his conduct, which was evidently prompted by reason.

"While a friend of mine was last week superintending his workmen in a wood, he observed his dog, a retriever, busily occupied in collecting mouthfuls of hay and withered grass, and carrying it all to one spot. On going to examine it, he found the deposit made was on a closely coiled hedgehog. The dog, having attained his evident purpose of rendering the spines harmless, proceeded to take up the heap with its contents, and then set off triumphantly toward home."

No human being could have acted in a more judicious manner; and had a man saved his fingers by enveloping the hedgehog in grass, he would not have felt particularly flattered if told that he had acted by instinct and not by reason.

A rather odd example of dog-reasoning occurred not long ago. A Newfoundland dog was walking with his mistresses, when he got into a quarrel with a costermonger's dog, fought him, conquered him, and left him howling on the ground. Seeing, however, that the animal would be in the way of the ladies, he returned, took up the animal in his mouth, and deposited him in the middle of the road, so as to allow them to pass without annoyance, and then returned to his usual position.

We will now pass to other animals.

We are often accustomed to use the name of ass as a synonym for stupidity, whereas it is one of the most intelligent animals in the world. The Rev. C. Otway has the following remarks on the subject:

"I assert that if you were to make yourself acquainted with asses, you would find them clever enough. I once purchased an ass for the amusement of my children. I did not allow him to be cudgeled, and he got something better to graze upon than thistles.

"Why, I found him more knave than fool; his very cleverness was my plague. My ass, like the king's fool, proved the ablest animal about the place, and, like others, having more wit than good manners, he was forever, not only going, but leading other cattle into mischief. There was not a gate about the place but he would open it; there was not a fence that he would not climb. Too often he awoke me of a summer's morning, braying with sheer wantonness in the middle of my field of wheat. I was obliged to part with him, and get a pony, merely because he was too cunning to be kept."

A correspondent of *Land and Water* gives an interesting account of a similar mode of proceeding on the part of two long-horned cows. The door of the hay-chamber opened outward, and was fastened by a latch lifted by the finger thrust through a hole in the door. The cows had seen this done, and, if left alone, would invariably open the door by inserting the tip of a horn into the finger-hole, lifting the latch, and then drawing the door toward them. He also describes the mode in which a cat opened a kitchen door, by jumping up and hanging on the handle of the latch.

Dr. Bell has recorded almost identical habits both of the horse and the cow, and I have heard similar stories in many places.

As if to illustrate this point still further, I have just received an account of a cow which could not be kept in the field, because she was in the habit of lifting the latch with her horn and then pushing the gate open.

The same correspondent mentions a horse which was accustomed to pump water for himself. The pump was in a corner of the horse-box in which the horse was shut for the night, and the coachman used to be puzzled at the fact that when he came in the morning the end of the stable was always an inch or so deep in water. At last he suspected that the horse might have been the delinquent, and so fastened him up without giving him any water, and watched him unobserved when let loose in the morning. The animal went at once to the pump, took the handle in his teeth, worked it up and down, and, when the water was in full flow, placed his mouth under the spout to drink. He could not endure being watched while pumping, and, if he saw any one observing him, would rush at him with open mouth in order to scare him away.

The mule, like the ass, is popularly thought to be a stupid and stubborn creature, and yet there are few animals more intelligent in their way. I can not resist relating one or two anecdotes, which are told by J. Froebel, in his work on South America. The mule, it appears, is a most difficult animal to manage, on account of its cunning. Force is of no use, and the Mexican mule-drivers pride themselves in their skill in managing the animals. At the end of the day's journey, the mules are unharnessed and allowed to go free, and are captured by the lasso when they are to be again harnessed. Some mules are so cunning, however, that even the experienced muleteers can scarcely capture them. Some of them assemble in a compact circle, with their heads all pressed together, so as to prevent the noose from settling on their necks, while others push their heads under the wagons or between the wheels. Others, still more cunning, stand still, and as the lasso rushes toward them, merely step aside and let it pass.

One mule, a white one, succeeded in baffling the attempts of the drivers throughout the whole of a long journey. As soon as the harness-time approached, it ran off for half a mile, and there stood until the whole train of wagons was in motion, when it quietly joined its companions. On one or two occasions it was captured by a couple of men on horseback; but it led them such a chase, wasted so much time, and fatigued the horses so much, that it got its own way and had a mere journey of pleasure, while all its companions were hard at work.

Another mule, which belonged to a convent, was equally averse to work. There were six mules, each being worked on one day of the week in regular order. This mule knew its own day perfectly well, and on that morning it always tried to keep the servants out of the yard by backing against the door.

The following account of a horse was sent to me by a clergyman:

"I had long entertained the idea of sending you a brief account of an instance of reason which occurred to my own knowledge, and indeed at our own door.

"A neighbor possessed a young foal, which, with his mother, used to pass our house daily, early in the morning, during our breakfast-time, and had a habit of straying upon a piece of waste ground which then occupied its front, but has since been inclosed and formed into a front garden. My daughter, who is extremely partial to horses, used to run out and offer the little animal a piece of bread.

"This went on regularly, until at last, when he was between two and three years old, he would not wait for the bread, but used to go to the door, plant his fore feet on the steps, so as to gain sufficient elevation, and then lift the knocker with his nose, afterward waiting for the expected morsel.

"Had I been a rich man, I would have bought him at almost any price; for his mother was a well-bred mare, and he promised to be a very useful roadster."

Here we have the two qualities of memory and reasoning displayed in a most unmistakable character. Indeed, if we suppose that a dumb man had acted as the horse did, we should have been disposed to marvel at the way in which human reason could communicate ideas without the aid of speech. In this case, the memory of the animal enabled him to expect his daily dole of bread, and his reason — not his instinct — taught him that, when the knocker was sounded, some one came to the door. It is evident that the horse had seen the knocker used, had noted the result, and had followed the example, using, of course, his nose in lieu of a hand.

Perhaps there are few of us who have possessed pet cats who have not seen the animals perform very similar feats. Although too small to reach a knocker or a latch, many a cat has been seen to knock at the door and to open it for herself, merely by jumping and striking the object with her paw.

Sometimes, when a door must be opened by means of a knob instead of a latch, the animal knows perfectly well that it is physically incapable of turning the handle, and therefore does not try to do so. But it will always find some way of intimating its wish to have the door opened, and will ask, as plainly as if it possessed speech, some human being to perform the task of which it feels itself incapable.

Some few years ago, Professor Cope related the proceedings of a tame monkey which he possessed—one of the common Capuchin monkeys.

The animal was kept in a cage, or rather was supposed to be kept in it, for he had a strong objection to confinement, and was sure to break loose sooner or later. He always directed his attention to the hinges, and no matter how firmly they were fixed, he was sure before long to extract the staples, pull out the nails, and so open the door at the hinges and not at the latch.

Finding that the cage could not hold him, his master had him confined by a strap fastened around his waist, after the manner of monkeys. The strap proved to be of no more use than the cage, for the crafty animal soon contrived to open it. This he did by the ingenious expedient of picking out the threads by which the strap was sewn to the buckles, and so rendering the fastenings useless.

Then he was replaced in the cage and carefully watched. Having rid himself of the strap, he thought that he might as well turn it to some useful purpose. So, having perceived that some food had fallen out of his reach, he took one end of the strap in his paw, flung the other over the morsel of food, and so drew it toward him. In this feat he displayed great accuracy of aim, seldom missing the object which he wanted.

Once or twice, when he had to make a longer throw than usual, he loosened his hold of the strap. The first time that this happened, some one handed him the poker. He took it, drew the strap toward him, and resumed its use as before.

Now I should think that no reasonable reader could deny that every one of these acts was prompted by reason, which, so far from being even aided by instinct, was acting in direct opposition to it. The instinct of an animal when confined or tethered in any way is to break loose by main strength, and the instinct of the monkey would have impelled him to force his way through the bars of the cage or to strain at the strap until he broke it. His reason, however, taught him to look for the weak part in both cage and strap, and, having found it, to work at that part alone until he succeeded in his object.

How was it possible for instinct to teach him that the hinges were the weak part of his cage, and that, if he could only manage to draw the staples or nails, the door would open and he would be free? How could instinct teach him that the stitches of the strap-buckles were to the strap what the staples and nails were to the hinges, and that, if he could pick out the threads, the fastenings of the strap would be rendered useless? Neither could instinct teach him to use the strap in the light of a lasso, or to employ the poker in regaining his lost weapon.

Baron Trenck himself could not have displayed more ingenuity in discovering the weak parts of his prison and bonds than did this little monkey, nor could he have shown more patience and perseverance in working at them. Indeed, there are many human beings who would not have been half as sensible.

CHAPTER IV.

REASON—(*concluded*).

Enjoyment of Work by Animals.—The Elephant and the Dray-horse.—The Miners' Horse.—"Jock" at his Work.—The New Forest Horses.—The Carrier's Horse and his Master.—Sir Hope Grant's Story of a Wounded Horse.—An Elephant in a Quicksand: Mode of Extrication.—The Cat and the Lobster.—Ingenuity of Rats.—Pigs *versus* Rats.—Crows *versus* Pheasants.—The Ravens and the Bird-traps.—Robbing the Railways.—My Raven "Grip" and his Run.—History of a Parrot.—A Self-tamed Red-breast, with his Well-educated Family; Death of the Father and Friendship of the Widow and Children.—The Cat "Patch" and the Mouse.—"Pret's" Mouse-chase.—Reason and Power of Combination in the Rat.—The Fox and the Grouse.—A Dog-strategist in Battle.

SOMETIMES animals take a pleasure in their work, and do it without needing any supervision. Elephants, as is well known, when once shown what their work is, will go on with it while their drivers are elsewhere engaged. Dray-horses may often be seen exercising their reasoning powers while drawing casks out of the cellars. The drayman in the cellar makes the rope fast, and calls to the horse. The animal understands the signal, and goes off with the rope, keeping an eye on the cellar-door. As soon as he has brought the cask safely to ground, he stops, backs to allow the rope to be removed, and then goes back for another cask. This may be seen almost any day in London.

Mr. J. Nelson Smith tells me that, while examining one of the American mines, he saw a horse which was doing his work without the assistance of any driver. As soon as his cart was filled with ore, one of the miners gave his signal, and the animal went off to the spot where his load was to be "dumped," waited until the cart was unloaded, and then returned for another load. The strangest point in his conduct was that he had to take a certain number of loads daily, and knew when his task was finished as well as did any of the men. Mr. Smith happened to be present at the time when he deposited his last load for the day, and, on seeing him trot off quickly in another direction, was told that he knew his work to be finished, and that he was going home, where he would meet a kind reception from his mistress.

A lady has sent me the following account of a horse of her own:

"We have an old horse named 'Jock,' a very wise beast, but cross-tempered. He fell when drawing Lord L.'s carriage, and, in consequence of his broken knees, was purchased cheaply.

"He knows his work so well that the man who accompanies the cart does not need to lead or drive him, Jock preferring to do his own work in his own way. I have often seen him take the cart to the exact spot intended, turn it round himself, and wait to have it loaded. When the cart is filled, he takes it to the spot where it is needed, and, after it is unloaded, brings it back again. He evidently enjoys the work, and seems to take a pride in it."

Horses will really do a wonderful amount of work without assistance, if properly managed, and will sometimes do so even when employed by owners who would scarcely be thought capable of acting as teachers. In the New Forest, a place tenanted by a race of human beings almost independent of their fellow-beings, and holding their laws and customs in equal scorn, I have often, when driving along one of the roads, been obliged to turn off the road, and to manœuvre both horse and vehicle into the underwood, in order to allow a train of wood-carts to pass. These wagons are constructed in cool defiance of the Act which prohibits more than a certain width between the wheels, so that a cart will occupy the full breadth of the road. No one drove the horses; but on each cart lay one or two men, utterly intoxicated, having managed to scramble into their vehicles under the knowledge that their horses would take them safely to their homes.

I know of a carrier's horse which acts in a similar manner, though not for a similar reason. The man has to make a night journey, beginning about midnight and ending about six A.M. The driver has such perfect confidence in his horse that he composes himself to sleep as soon as he has started, knowing that the animal will stop at the right house. Sometimes he is asleep when the journey is over. The horse, after looking around at his master, and seeing that the stopping of the cart has not aroused him, begins to stamp on the ground, and rattles his harness until he awakes.

General Sir Hope Grant, in his diary of the "Incidents in the Sepoy War," narrates a most remarkable instance of reason on the part of a horse :

During the war, after the Secundra Bagh had been taken by our troops, the Europeans were aroused by musketry from some unseen quarter. Sir Hope's nephew then went to the place, gave his horse to a Sikh soldier to hold, and went inside, when he found that some of the rebel sepoys were on the top of the wall. Finding themselves discovered, the men, with the curious indifference to life that characterizes their race, came down and were shot.

Suddenly a loud explosion was heard. An awkward soldier had fired into a barrel of powder, which, together with a quantity of loose powder that was scattered about, exploded, and did much damage. The non-commissioned officer in charge of the Sikh party was so severely burned that he died a few days afterward; and several were killed, among whom was the man who was holding the horse.

The animal was so scorched that he had to be shot. It so happened that the man to whom this task was intrusted aimed badly, and, instead of killing the poor creature, only inflicted a severe wound in the head.

The horse broke away, galloped directly toward a picket of the enemy, dashed through them in spite of their fire, and was soon out of sight. Next morning it was discovered that the horse had made his way five miles in a straight line, and had gone direct to the sick-horse stables of the 9th Lancers. In fact, he had acted exactly as a wounded soldier would have done—gone to the hospital and reported himself sick.

I wish I could give a more pleasing end to the story, but the poor horse was found to be so fearfully injured that the most humane course was to destroy him at once.

A very similar exercise of reason was displayed by a little Welsh pony.

At Rhyll there are many of these animals let for temporary hire, and among them was one that was ridden by a young lady in delicate health, who was obliged to keep to a very slow pace. One day in the year 1873 the pony was seen dashing along at full gallop, until it reached a blacksmith's forge, into which it went without a pause, carrying its unwilling rider with it.

The astonished blacksmith tried to lead the animal out of the forge, but it resisted this strongly, and he found that it had cast a shoe, which it wished to have replaced. Now in this instance, as in many others, reason conquered instinct. The instinctive feelings of a horse are strongly opposed to the operation of shoeing, and some horses can scarcely ever be made to stand still under the process. It is very natural that they should not like their feet to be hammered and filed and scraped and scorched, and it therefore requires a very determined exercise of reason to induce an animal voluntarily to counteract its own instincts.

In the following account of an elephant's ingenuity in extricating himself from a quicksand, instinct is shown to have no part. The story was sent to me by the gentleman who witnessed the occurrence, and was one of the party.

"It was at the close of a 'pig-sticking' meet on a large island in the Ganges, opposite Cawnpore, in June, 1873, that an event occurred which excited my admiration. With three friends I had been riding hard all the afternoon, and, feeling very tired, we determined to go home on the elephant.

"We had traveled some way, and were nearing the river, when one of us noticed that the ground looked rather unsafe, and that the elephant seemed to become uneasy. So we all decided to dismount and walk to the river, previously instructing the mahout to take the elephant by a short circuit, so as to avoid the soft ground. The man, however, evidently considered that he knew better than we did; and we therefore went straight on, not thinking of looking around.

"We had not proceeded far when we heard the elephant trumpeting (a well-known signal of distress or anger), and on looking around saw that the poor beast was in a quicksand, and that the mahout had dismounted and was making the best of his way out of reach of the animal. I may here mention that to be on an elephant's back or within his reach under such circumstances is certain death : he is sure to take hold of a man and place him under his feet, so anxious is he to get something solid to stand upon.

"We were then about fifty yards from the river, and it was rapidly getting dark. The elephant was making frantic efforts to escape out of his difficulties, and the ground heaved all around him. How to help him we did not know; for he was sinking deeper and deeper, and go near him we dared not.

"As good fortune would have it, there happened to be at hand a number of large planks which had been left by some villagers. We went as near the elephant as we dared, and threw the planks within his reach. The clever animal seized them in his trunk, drew them to him, and laid them one upon the other in front of him. When he thought that he had enough, with one gigantic effort he got his fore legs out of the quicksand, and in a short time he had managed to extricate himself, and was standing safely on the planks, though trembling all over.

"He had still some fifty yards to go before he could reach the river, and the intelligent beast never moved a step until he got a plank and placed it in front of him. He thus moved on, step by step, on successive planks, until he reached the river. The mahout then remounted him, and he crossed the Ganges in safety. This was no small relief to our feelings; for the loss of the elephant would not only have been a very costly business, but the mode of his death would have been inexpressibly painful."

The following account of reasoning in a cat was communicated to me by its mistress, Lady E., whom I have known for many years. The animal evidently felt surprised that such a thing as an empty plate should be allowed upon a breakfast-table, and so, in her own way, showed her mistress how a plate ought to be filled.

"Our breakfast-room had bow-windows, and the houses were very near each other.

"One morning, when the windows of both houses were open, our younger cat, Tiny, disappeared into our neighbor's window, and a few minutes after rushed back into our room, and, leaping upon the breakfast-table with a lobster in her mouth, held it over an empty plate. She evidently only wished us to see it, as she would not allow any one to touch it, and darting out of the window again, with the lobster still in her mouth, she replaced it upon the table without taking any, and came back to our room.

"The lobster was returned so carefully that our neighbors assured us they should not have known it had been touched."

The same lady has sent me several anecdotes of this same cat and her mother "Rosie," all of which are interesting, and serve admirably to illustrate the subject of this work.

Several good instances of reasoning as displayed by rats are given in Hardwicke's *Science Gossip* for July, 1871. A number of rats had got into a basket of grapes, and devoured a considerable part of the contents. The man who discovered them replaced the basket, in hopes that they would again visit it and be caught; but the wary animals never again came to the basket in which they had been detected.

They were so numerous and so bold that they used to come and pick up the crumbs from between the men's feet as they sat at meals. "Wishing for a shot at some of them, I dropped a few grains of maize on the ground, and took up my position, gun in hand. Soon one rat bounded across the space as if in great alarm; but no rat touched a grain of the corn, which was exposed for several days and nights, being at last crushed and lost by the passing of feet and goods.

"Rats were numerous in the pigsties, and ate with the pigs, one of which I turned out of her sty, and contrived a trap-door to close the trough by pulling a cord. I baited the trough with ground maize, of which they are very fond; but neither by day nor by night would a rat venture there as long as the pig was excluded. Returning the pig to the sty, the rats also returned."

I know of a similar case in which the rats were so many and so bold that they forced themselves into the troughs at feeding-time, would not be driven away, and consumed no small amount of the food which ought to have gone to the pigs. The owner of the pigs then laid a gun so as to rake the trough, turned out the pigs, and had the trough filled as usual. Not a rat would make its appearance; and at last the pigs were put back, when the rats came trooping in as numerous and as bold as ever.

Now, in these cases, the rats could not have known the precise danger which menaced them; but they saw that something unusual had happened, and therefore inferred that it would be the safer plan to keep out of the way until the ordinary conditions were restored.

Many birds display great reasoning powers, and act in a way that would do credit to any human being. From the many anecdotes which have been placed at my disposal I select only a few, none of which have as yet been published.

In places where pheasants are preserved it is customary to give them their food in such a way that other birds can not get at it. This is done by placing it in a feeding-box, which is closed by a lid, communicating by a lever with a perch.

The weight of the lid is so adjusted that when a pheasant stands on the perch the lid is raised, and the bird can get at the food. The pheasants soon learn the object of the perch, for, when these boxes are first introduced, a few beans are laid on the outside of the lid. The bird gets on the perch in order to reach them, and so exposes the stores of food in the box.

Such an arrangement is made at Mountquharrie, Cupar, Fife; and one day a gentleman was watching the pheasants and their boxes on the lawn just before the house, and saw a crow also watching them. Presently the crow flew to one of the boxes, settled upon the perch, and expected the box to open. The bird, however, being much lighter than a pheasant, was unable to lift the lid in spite of all its efforts. After several ineffectual attempts it flew off to a tree where there was another crow, and a grand jabbering ensued. The two crows then flew to the feeding-box, both settled on the perch, and their united weight was sufficient to raise the lid.

It is impossible to attribute this proceeding to any thing but reason. Instinct is wholly out of the question in such a case as this. The bird first watches the pheasants, and learns that by settling on a certain perch the box is opened and the contents attainable. It then proceeds to follow the example of the pheasants, judging that the same result would follow. Finding that, although it acted exactly as did the pheasant, the lid was not raised, it set itself to discover the cause of failure, and, as we have seen, succeeded in so doing. Having reflected that the pheasant could lift the lid on account of its superior weight, the bird calculated that two crows might be equal in weight to one pheasant. So it goes off to find a comrade, explains the state of things in its own bird language, and the two then co-operate in producing the desired effect. No human being could reason more correctly, or reduce its theory to action more successfully.

That the raven can act in a similar manner is shown by an anecdote sent by Mr. R. Ball to Mr. Thompson, and quoted in his "Natural History of Ireland:"

"When I was a boy at school, a tame raven was very attentive in watching our cribs or bird-traps, and when a bird was taken he endeavored to catch it by turning up the crib; but in so doing the bird always escaped, as he could not let go the crib in time to seize it. After several vain attempts of this kind, the raven, seeing another bird caught, instead of going at once to the crib, went to another tame raven and induced it to accompany him, when the one lifted up the crib and the other bore the poor captive off in triumph."

Crows are wonderfully sagacious, and seem to notice every thing.

A gentleman, one of the principals in a well-known engineering firm, tells me that the way in which crows rob the railway-boxes of the grease is quite notorious among those who are connected with the lines.

As my readers are probably aware, each of the wheels has an iron box over the axle in order to contain the grease which lubricates the wheels. Cocoa-nut oil is used for this purpose, as it is solid at moderate temperatures, and only melts and sinks upon the axle when the latter is heated by over-friction. Indeed, if cocoa-nut oil had not been discovered, it is difficult to imagine how railways could be carried on. The boxes are closed with spring lids, and we have most of us seen the porter, armed with a little pail of cocoa-nut oil and a wooden spatula, open the box with the spatula, fill it with the yellow grease, and slap down the lid upon the box, where it is kept in position by a spring. This is absolutely necessary in order to prevent the oil from being mixed with the cinders ejected from the engine and the particles of earth driven up by the wheels.

Now it happens that crows value the cocoa-nut oil as much as we do, but for a different reason. They consider it to be a great dainty, and so, when a train is standing still on a siding and no one near, the crows flock to it, substitute their strong beaks for the porter's wooden spatula, pry up the spring lids, and help themselves to the yellow oil.

It is evident that they must act from reason and not from instinct. Some of them had seen the porters lifting up the lids, and had followed their example. All the crow tribe are wonderfully expert in the use of their beaks, and the dainty manner in which a raven, a magpie, or a jackdaw will turn over, twist, and display with its beak any object that may excite its curiosity could scarcely be surpassed if the bird possessed a hand instead of a beak.

My raven, "Grip," who unfortunately died from eating too much linen, had astonishing delicacy in the touch of his great iron beak. If I tied a knot in a piece of string and left it within his reach, he was sure to untie it, and then walk about triumphantly with one end of the string in his beak. He had a large wooden cage made from a chest, and faced with strong iron bars. A hole was cut in the end of the box, leading to a large "run," inclosed with wire netting.

There was not a spot at which the netting had been joined that had not been tested by Grip's beak, and more than once I have just been in time to prevent his escape. He always resented my interference, and used to seize in his beak the wire with which I was making the defect good, and try to pull it out of my hands. At last he gave up the wire net, and turned his attention to the bars of the cage. They were much too strong for him to bend, but he deliberately set to work at one of the central bars, and dug away the wood in which it was set until he had loosened it at the bottom. Fortunately I was just in time to see him pulling at the bar, or there would have been an escaped raven and frightful havoc among the poultry kept by my next-door neighbor.

Directly Grip saw me he set up a great squall, and did his best to get out the bar before I could reach him. I at once sent for wire and pliers, and at last succeeded in connecting the whole of the bars with cross-wire, so that unless all the bars were dug out both above and below they would hold their place.

Grip was horribly angry during the time, and tried to annoy me as much as possible by striking at my fingers through the bars, and trying to pull away the wire. Once he did seize the pliers, and I was obliged to bring on the scene my dog "Bosco," whom Grip hated beyond conception, before I could induce him to drop the pliers. Bosco's presence, however, elicited a scream of rage; and as the pliers fell from his beak, I secured possession of them. He afterward tested the wires from end to end, tried to undo every knot, and, finding himself baffled, gave up the whole business as a bad job.

Here are some parrot anecdotes, all perfectly original:

"A parrot, belonging to one of our servants, very soon knew us by name, and could distinguish the tread of its favorites, showing its joy by ruffling its feathers and making an odd noise in the throat. 'Polly' was very tame, and was sometimes allowed to walk about the house, always announcing its arrival in a room by 'Polly going a-walking.' In hot weather she enjoyed having water poured over her, and when satisfied would say, 'That's enough.'

"She used to tease our large dog by whistling loudly, and calling him 'Bran! Bran!' on which he ran in and looked around, and on the cook coming in, Polly would say reprovingly, 'Go back, Bran, go back;' out went Bran, and by and by, when the cook's back was turned, the same scene was acted over again, until Bran grew wiser and neglected the call.

"Polly was a very accomplished bird, and, when quite alone, could be heard going through her acquirements. She sang 'Cheer boys, cheer,' very plainly, and could dance. If any stranger went into the kitchen, and no one was there, Polly called out, 'Somebody's wanted;' and she has more than once startled people by saying, 'What's your business?'

"We used to go in and see Polly before we went to bed, and she always said 'Good-night' several times, each time in a different tone of voice. She called mamma 'my dear' until told that it was not respectful, after which she always said 'ma'am.' The remarks this bird made were so apposite that it really seemed at times as if it understood what was going on."

I know a parrot, or, correctly speaking, a ringed parrakeet, that acts, as the servants say, "just like a Christian." If told to call the cat, she will sometimes mew loudly, and sometimes call the cat by its name, "Winks," which is an abbreviation of Tiddlywinks. She makes the room ring again with the name, her voice is so powerful. Sometimes she will play at hide and seek; and if her mistress gets under the table, Polly traverses it in all directions, and, not seeing her, knocks violently on the table with her beak, in order to induce her mistress to come out of her hiding-place.

In the following history of a self-tamed redbreast, we shall see that instinct plays but a very small part, almost the whole of the bird's proceedings, as well as those of his family, being instigated by pure reason without any admixture of instinct. To the lady who sent me the anecdote I am indebted for several of my most interesting accounts of animal life. She does not wish her name to be mentioned, but it is well known throughout the whole literary world:

"In the years 1864 and 1865 a robin made itself at home in my dining-room, always coming to the window and tapping to have it opened at breakfast-time. When he came in, he shared my oatmeal porridge with me, seating himself on the edge of the cup and picking out such grains as caught his fancy. He then picked up crumbs of bread or toast, and, when he had satisfied himself, he sat on the back of my chair and sang, or sometimes betook himself to the top of a large screen. When he wished the window to be opened for him, he used to make a peculiar little noise, unlike any sound I ever heard from a bird—not loud, but very much like articulate language.

"As you may fancy, he was a great favorite with every one in the house. If the day were very cold, he always seated himself on the edge of the fender as soon as he was let in, puffing out his feathers to receive the heat, and, when he found that he was warm enough, he came to his breakfast.

"During the summer of 1864 he came occasionally to the window, but seldom came in, and then only for a moment, though he would sometimes follow me out of doors. In the winter of 1864-5 he again established himself in the house, on his own familiar terms, and became even a greater pet than ever. He then began to prefer the butter-cooler to the porridge-cup for his breakfast, but I never allowed him to take too much. He almost lived in the house, sometimes remaining all night when the weather was bad.

"When summer came around again, he appeared one day at the window with his wife and children, who sat on the ledge of the window while he entered and took food out to them. It then came out that of late he had often been detected in carrying off food from the peacock's bowl which I kept in the dining-room; this food he had doubtless carried to his lady in her nest: the dining-room window, being mostly open in summer, gave him access to the bowl.

"A sister-in-law of mine and her daughters came to stay with me just then, and to see the little redbreasts get their breakfast daily from their papa was one of our morning's amusements.

"But, alas! one day he came looking very ill, with his feathers puffed out, and looking twice his natural size. I observed that he swallowed large lumps of butter himself while helping his young ones. This went on for some days, and at last he did not make his appearance at all; his wife and family came without him, and then we knew that he must be dead. There was general mourning for poor 'Bobby' in the house. I have never had so tame a redbreast before or since, though his wife and children, who seemed to miss him much, still continued to receive their dole at the window.

"I heard a still more wonderful story about a robin from my sister-in-law, who knew the lady to whom the bird belonged. She had made it so tame that it used to fly after her carriage; and when she went in the winter to spend a few days with a friend who lived several miles from her house, the bird followed her. On the following morning, when she opened the window according to custom and called the robin, he at once entered the room and perched on her finger.

"Was not this very like reason? It certainly was a combination of ideas. The bird had followed his mistress to a strange place, slept there, and came at her call, trusting to her for his breakfast. My sister-in-law was staying at the house at the time, and witnessed the circumstance."

In the former of these two cases, reason taught the bird to conquer its instinct, which teaches it to fear man and avoid him. The bird soon found that he was being kindly treated, and, reasoning upon such premises, came to the conclusion that he would be treated in the same manner for the future. Then, that birds must have a language in which to express their ideas is evident from the fact that his wife and family accompanied him to the house, and waited outside while he went and brought out food for them. The reason why they did not enter the house is evident to all who know the habits of the redbreast. It is one of the most jealous of birds, and never will allow another bird to enter the place of which it has pleased him to consider himself the owner. There can be little doubt but that he had previously forbidden his family to enter the house where he felt himself a privileged inmate.

The capability of cats for opening doors, ringing bells, etc., is perfectly well known. There was a cat named "Patch" who was a great adept in these arts. One evening she came out of a bedroom in a state of great excitement as the occupant went in, mewed and fidgeted about; went up to an unlighted candle, though there was a fire in the room, back to the lady and then again to the candle, and would not be contented until it was lighted. Then she drew particular attention to the window-curtain, reaching up with her paw as far as she could, and touching it. The curtain being shaken, out dropped a mouse, which Patch immediately seized and carried off. She had, probably, previously brought it into the room, as she was in the habit of doing so with her prey, and on two or three occasions dead mice were found deposited in the bed.

My own cat, "Pret," has often behaved in a similar manner, and has brought me to help him in getting at a mouse which had hidden itself in some spot where he could not reach it.

I might multiply anecdotes to an indefinite extent, but have thought it better to take a comparative few, nearly all of which have been as yet unpublished. The reader will see that in no one of these cases does instinct play any part, and that in the generality of them the reasoning powers of the animal have overcome its natural instincts.

Here is an example of reason and the power of combination in the rat. The writer was at the time resident in Liverpool:

"In my garden there is a conservatory, along the roof of which is trained a vine, on which the fruit would not ripen for the last few years, so I had the vine inclosed in a glass frame in the hope that, the heat being confined, the grapes would ripen better than when exposed to the cold night air. This plan being successful, I had this year a plentiful crop of large-sized bunches of grapes. These, however, began to disappear very quickly as soon as ripe, but not bunch by bunch as would be done by thieves, but only the ripest grapes of each bunch were taken.

"At first, I thought that some of the boys working in the garden had been helping themselves; but all denied it, and no one had seen them near the glass house. Then I sealed up the door of the covering, but still the fruit disappeared. So I told the gardener to cut all the good fruit and take it into the house when I returned home in the evening; after giving the order, the gardener came in with gleeful visage and said, 'I've got the thieves, sir,' and told his tale in that roundabout way which men in his condition love, of which the following is the condensed description:

"'When lying on my back for rest after cutting a lot of branches, I heard a scuffling sort of sound, and looked around and saw five or six large brown rats come into the frame; they then jumped up at the lowest hanging branches and managed to knock down two or three grapes, which they proceeded to eat like a squirrel, sitting up on their hind legs and holding the fruit in their front paws.

"'Soon after, a large female, followed by four young ones, came in; and the old one ran up the vine and bit off one of the ripest bunches, which fell down to the expecting young ones below, who fastened on it and began to eat. Then,' concluded the old man, 'I could not keep my laugh any longer, but shouted out, which sent them all head over heels out, as if a dog were after them.'"

A curious instance of reason in the fox has been furnished to me by an eye-witness:

"I will now tell you a story of a fox. Some years ago, when I lived in a lonely but beautiful part of the Lammermoors, there came a dreadful snow-storm. All nature was white for miles, as if wrapped in a winding-sheet, and birds and beasts were put to strange shifts for food.

"I was talking with one of my shepherds, when far away on the opposite side, and on the top of what is here called a cleugh or hollow, I espied a small dark object. It was the only one in the vast expanse of snow, and it appeared to me to be moving. I pointed it out to the shepherd, who said that it was a tuft of heather, from which the snow had drifted. I watched it more carefully, and, feeling sure that it really did move, I went into the house for my gun, and told the shepherd to accompany me.

"Slowly we plodded our weary way through snow up to our waists in some places; and when we arrived within a few hundred yards of the mysterious object, it was revealed in the shape of a crafty fox, who deliberately walked away, every now and then stopping to look at us.

"It was evident what he had been doing. He had coiled himself round so as to look like a bunch of heather (and done it so well that he had even deceived the practiced eyes of the shepherd), and thus decoyed the hungry grouse near enough to seize them. That he had succeeded was plain, from the feathers and other remains of several birds which lay near the spot where we first saw him. Foxie is a rare purveyor, and nothing can beat him."

A rather amusing instance of reason in a dog has been narrated to me. The animal was a Newfoundland, and of a quiet disposition. There was, however, a much larger and quarrelsome dog of the same kind, who was frequently meeting "Lion," and taking every opportunity of molesting him.

One day the big dog met him, and evidently bent upon a fight. Whereupon Lion, knowing that he was no match for his antagonist without some aid, ran off to a neighboring manure-heap, and rolled himself over and over in it, until he was completely covered. He then went back to his enemy, challenged him, fought him and beat him thoroughly, and after that victory the big dog always gave Lion a wide berth.

CHAPTER V.

LANGUAGE [OF ANIMALS].

Ideas Useless unless they can be Transmitted.—Language the Means of Transmission.—Various Kinds of Language.—The Spoken Language, or Language of Words.—The Gesture-Language, or Language of Signs.—The Language of the Eye, or a Direct Transmission of Ideas without the Aid of Words or Gestures.—Language of Insects.—The Wasps at my Breakfast-table: a Messenger and Result of the Message.—Language among the Ants: Severity of their Military Discipline.—Ant-Undertakers.—A Summary Execution.—Power of Combination and Submission to a Single Leader.—Comparison with the Egyptian and Assyrian Laborers.—Language among Dogs.—A Tempter and his Victim.—Language and Combination among Dogs.—Ditto among Wolves.—A Specific and a Universal Language among Animals.—Language and Combination among Baboons.—Monkeys and the Charge through the Mud.—Division of Labor between Dogs.—Mutual Arrangements between a Dog and a Cat.—Rook Parliament seen by a Lady in England.—Ditto by a Gentleman in India.—Ditto by a Gentleman in Cornwall.—A Thrush Parliament Discussing the Fruit Question.—Martins Sitting in Judgment on a Sparrow, and Killing Him.—"Beau" and his Rescuer.—A Quarrel and a Peacemaker.—The Goose, the Ducklings, and the Hen.

The possession of ideas, whether they be right or wrong, infers more or less reason in those beings who possess them. Those ideas would be absolutely unknown without some means of transmitting them, and such means we call by the name of Language.

There are several kinds and degrees of language known to ourselves. First comes the spoken language, in which ideas are clothed in certain definitely regulated sounds. Then there is the written language, in which those sounds are reduced to form, and are heard with the eye instead of the ear.

Then there is the language of gesture, which is little employed among ourselves, but in some parts of the earth forms a necessary concomitant to the spoken language, or can be substituted for it. The Bosjesmans of Southern Africa, for example, are unable to converse with freedom when in the dark, the visible gestures being needed to supplement the audible words. This necessity is so great that if they wish to talk in a dark night they are obliged to light a fire.

Among the North American Indian tribes the language of gesture forms an important part of every man's education. There are very many of these tribes, and they all speak different dialects, which in many cases vary so much that they are practically different languages.

Were it not for some other means of communication besides spoken words, no one would be able to converse with another who did not happen to belong to his own tribe. Gestures, however, take the place of words, and form a universal language. This sign-language is very simple, is based upon definite principles, and is easy of attainment. Captain Burton has written an account of the sign-language, which ought to be carefully read by all travelers. The language as given by him is easily mastered, and by its use, acquired in a few hours, an Englishman would be capable of conversing with any of the savage tribes of North American Indians without understanding a single word of their spoken language.

The English, in consequence of their physical constitution, which their Continental neighbors are pleased to call "phlegmatic," use gesture-language less than almost any nation upon earth, looking upon gesture in connection with language much as they do upon ornament in connection with objects of utility. Yet even they use it, though sparingly, and almost unconsciously.

That its use is natural is shown by the untaught and graceful gesture-language of a child, which is able to express its thoughts by gesture long before it obtains the power of speech. I knew a child who managed to express himself so well by gesture that he did not trouble himself to speak a word until after he had completed his third year. His mother was terribly distressed at his backwardness; but after he found the use of his tongue he more than compensated for his previous silence, and I fancy that his mother would occasionally have preferred an interval of the gesture-language which had been so distasteful to her.

In maturer years this silent language survives.

To take a few familiar examples: The uplifted finger expresses the idea of warning as plainly as if the word had been used. If one person tell another a tale, and his narrative be received with an almost imperceptible shrug of the shoulder, incredulity is expressed as clearly and as offensively as if the lie had been given in words. Similarly, the upraised eyebrows express wonder, but at the same time imply belief.

To shake the closed fist expresses menace, and indeed such a gesture is actionable at law. To present the palms of the hands toward an object expresses rejection, while the open arms equally express acceptance. There are some ladies who are addicted to the feminine vice of tossing their heads when they meet with any thing which does not happen to suit them at the moment. It is really wonderful to see how much they enjoy it, and how they think themselves to have elevated their dignity together with their noses above the ordinary level of humanity. Their idea is a ludicrously false one, but they certainly express it by their gesture.

Again, words can not express contempt more forcibly than the action of snapping the fingers or turning the back; nor can words be more expressive of veneration than the act of bending the knee. Words are not needed to express devotion when the clasped hands and uplifted eye are seen; while remorse is shown by the cowering form crouching to the earth as if crushed by the weight of guilt, and conscious innocence by the erect body and uplifted head.

Not to multiply further examples which will strike any one who takes the trouble to think on the subject, it is evident that ideas can be conveyed by gestures without the use of words, and that any mode of transmitting ideas is a form of language.

The gesture-language is that which is chiefly used by the lower animals when they wish to convey their ideas to man, and, in its way, it is as perfect a language as that which was employed by the child above mentioned, who did not choose to take the trouble of speaking when he could make himself understood by gesture; and, whether these gestures be used by man, child, or beast, they are intended for the transmission of ideas, which are the result of reason, and not of instinct.

Painters would be in a very bad way if they were not aided by the natural language of gesture. They can not paint ideas, but they can paint the gestures which are expressive of ideas, and so can make themselves as well understood as if they had made use of the written language. Indeed, the same model does duty for all kinds of personages and all kinds of emotions, as long as the gestures can be represented. An old, gray-headed, long-bearded man, with his hair tossing in the wind and his hands wildly clinched, represents grief and madness, as personated in Lear. The same individual, with face upraised and a harp on his knee, will be adoration, personified by David. Let him shut his eyes and hold out his hands, and he represents dignified penury in the person of Belisarius. The same rule holds good with sculptors. Man really could not go through existence without a gesture-language, and that language, as we shall presently see, is the common property of himself and the lower animals.

Even among ourselves there is a recognized language of signs, namely, that by which we can exchange ideas with the deaf and dumb. It has been reduced to a form almost as definite as the written or spoken language; and it is worthy of notice that very many of the signs are identical with those in use among the Indian tribes. Thus a deaf-and-dumb man who had learned the sign-language would be able to converse with the Indian tribes; while a man who was in possession of his powers of speech and hearing could neither understand them nor make himself intelligible to them if he were ignorant of this simple code of signs. I have seen evidence taken in a court of law by means of the sign-language, and such evidence was accepted as if it had been spoken or written.

Lastly, there is the language of the eye, by which ideas are interchanged without the necessity of words or gestures. It is essentially the language of idea, and by it spirit speaks directly to spirit, conveying by a single glance of the eye thoughts which whole volumes would fail to express.

There is none so obtuse that he can not understand the fiery glare of anger, the soft, beaming glance of love, or the dull, purposeless stare of hopeless sorrow. When the mother contemplates her infant, her entire soul is poured through her eyes, and no language is adequate to express the boundless love which is manifested by the eye alone.

The look of appeal is sufficiently recognizable to be expressed by the painter's art, an admirable and familiar example of which is seen in the two faces in Millais's "Huguenots." Solemn question and equally solemn response can be given in a moment, and without the use of word or sign; and there are those who have known a single glance given and returned change the whole course of two lives.

If animals possess reason in common with man, it is evident that they must be able to interchange thoughts with each other and with man, when brought in contact with him. They must possess a language of some sort, by means of which they can understand each other, can comprehend human language, and render themselves intelligible to man. All these conditions are fulfilled in the lower animals, and the inference to be drawn from them is self-evident.

There is one distinction between the capability of understanding their own language and that of man, namely, that they are born with the one and have to learn the other. Newly hatched chickens, for example, understand their mother perfectly well, though they have only entered the world an hour or so ago; they know what she means when she calls them to find what she has scratched up for them, and they know what to do when she gives them warning of danger. They, again, are able to talk to their mother, and even the most incurious must have noticed how different are their tones under various circumstances—say, for example, the little piping notes of content when all is going on well, and the cry of alarm when they have lost their way or are otherwise frightened.

Looking at the nervous system of insects, in whom there is no definite brain, but merely a succession of ganglia united by a double nervous cord, many physiologists have thought that reason could not be one of the attributes of the insect race. Yet nothing is more certain than that they are able to converse with each other and communicate ideas, this fact showing that they must possess reason. As far as we know, the hymenopterous insects—namely, the bees, wasps, and ants—are the best linguists of the insect race, their language being chiefly conducted by means of their antennæ. A good example of this was witnessed by me in the summer of 1872.

At breakfast-time some pieces of the white of an egg were left on a plate. A wasp came in at the window, and, after flying about for a while, alighted on the plate, went to the piece of egg, and tried to carry it off. Wishing to see what the insect would do, I would not allow it to be disturbed. After several unavailing attempts to lift the piece of egg, the wasp left it and flew out of the window. Presently *two* wasps came in, flew direct to the plate, picked up the piece of egg, and in some way or other contrived to get it out of the window. These were evidently the first wasp and a companion whom it had brought to help it.

I had a kind of suspicion that when the wasps reached their home they would tell their companions of their good fortune, and so I put some more egg on the plate and waited. In a very short time wasp after wasp came in, went to the plate without hesitation, and carried off a piece of egg. The stream of wasps was so regular that I was able to trace them to their nest, which was in a lane about half a mile from my house.

The insect had evidently reasoned with itself that, although the piece of egg was too heavy for one wasp, it might be carried by two; so it went off to find a companion, told it the state of things, and induced it to help it in carrying off the coveted morsel. Then the two had evidently told the other inhabitants of the nest that there was a supply of new and dainty food within reach, and had acted as guides to the locality. Here is positive proof that these insects possess a very definite language of their own, for it is impossible that human beings could have acted in a more rational manner.

Every one knows that wasps carry out one of the first principles of the military art by always having the gate of their fortress guarded by a sentinel. Should there be danger, the sentinel gives the alarm, and out dash all the inhabitants at the offender indicated by the sentinel.

It is clear that, out of the many hundred wasps which form a full-sized nest, the individual who is to act as sentinel must be selected, and its task appointed. We do not know how the selection is made, but that such is the case is evident; for the rest of the wasps acknowledge their sentinel, trust to it for guarding the approaches of the nest, while they go about their usual task of collecting food for the young and new material for the nest.

As for the ants, some of their performances are absolutely startling, so closely do they resemble the customs of human civilization.

They have armies commanded by officers, who issue their orders, insist upon obedience, and on the march will not permit any of the privates to stray from the ranks. There are some ants which till the ground, weed it, plant the particular grain on which they feed, cut it when ripe, and store it away in their subterranean granaries. There are ants which are as arrant slaveholders as any people on earth ever were. They make systematic raids on the nests of other ants, carry off the yet unhatched cocoons, and rear them in their own nests to be their servants.

There are ants which bury their dead—a fact which was discovered by accident.

A lady had been obliged to kill some ants, the bodies of which lay about on the ground. Presently a single ant found its dead companions, and examined them and went off. Presently it returned with a number of others, and proceeded to the dead bodies. Four ants went to each corpse, two lifting it and the other two following —the main body, some two hundred in number, following behind. The four bearers took their office in turns, one pair relieving the other when they were tired. They went straight to a sandy hillock, and there the bearers put down their burdens, and the others immediately began to dig holes. A dead ant was then placed in each grave and the soil filled in. The most curious part of the proceedings was that some six or seven ants refused to assist in grave-digging. Upon which the rest set on them, killed them, dug one large hole, and tumbled them unceremoniously into it.

In Froebel's work on South America there is a good account of the proceedings of some ants:

"I had several opportunities of observing the manners of several kinds of ants living in the houses. All of them are very inoffensive and even useful creatures. On one occasion I witnessed a remarkable instance of the concerted and organized action of a crowd of them. They were of a minute species, but, by the wonderful order and speediness with which they worked together, and which it would have been difficult to realize with men, they succeeded in performing a task apparently quite beyond their capability.

"They carried a dead scorpion, of full-grown size, up the wall of our room, from the floor to the ceiling, and thence along the under surface of a beam to a considerable distance, when at last they brought it safely into their nest in the interior of the wood. During the latter part of this achievement they had to bear the whole weight of the scorpion, together with their own, in their inverted position, and in this way to move along the beam.

"The order was so perfect that not the slightest deviation from an absolute symmetry and equality of distances and arrangement was observable in the manner of taking hold of the body of the scorpion, and in the movement of the little army of workmen. No corps of engineers could be drilled to a more absolute perfection in the performance of a mechanical task. According to a rough calculation, there must have been from five to six hundred of these intelligent little creatures at work. Besides those engaged in the transport, none were seen. A single one was sitting on the sting at the end of the scorpion's tail, as if placed there to overlook and direct the whole movements; all the rest were, without exception, at work. The operation may have lasted about an hour."

This scene is an exact reproduction, in the insect world, of the manner in which the ancient Egyptians and Assyrians conveyed their colossal statues to their places. There we see hundreds of men all dragging at the multitudinous ropes attached to the car on which the statue lay, and all pulling in time to the gestures of a single man placed on the top of the statue. The ants, however, had a still more difficult task than the men; for they possessed no carriage on which to lay the scorpion, and were obliged to sustain the whole of its weight as they passed over the ceiling.

In the same work, Froebel has narrated another example of the manner in which ants can combine, and make themselves intelligible to their fellow-insects:

"Another time I witnessed the transmigration of a whole state or commonwealth of ants, from a hole in the wall, across our veranda, into another hole in the opposite wall.

"Two facts struck my attention in this case. The first was, that the marching army of these insects, all moving in one direction, consisted of individuals of such a difference in size and shape, that to consider them as belonging to one species seemed very difficult, and the idea of a commonwealth of different insect nationalities was strongly suggested.

"The second was, that some little beetles, of the family of *Coccinellidæ*, marched along with the ants from one hole into the other; not quite of their own will, for I observed that several times one of them tried to deviate from the line, but was quickly brought back to the ranks by some of the ants placing themselves at its side. The fact of little beetles, of the very family just mentioned, existing in the nests of ants is well known; but it is of considerable interest to see the fact repeated in distinct climates, with different species of insects of both tribes, and under opposite circumstances."

As to the different sizes of the ants, all entomologists know that, in the hotter parts of the world, the males, females, soldiers, and workers of the same species will vary in size from that of a wasp to that of a common garden ant, and that the shape and aspect are as different as their size. The second point is a very curious one. It has long been known that many beetles live in ants' nests, but I believe that this is the only record of the beetles accompanying the ants in their migrations.

We will now proceed to some of the higher animals.

The Scotch shepherds, who are brought into constant companionship with their dogs, fully believe that the animals not only understand the words of their masters, but have a language of their own in which they can communicate ideas to each other. So certain are they of this that a shepherd is quite as fastidious about his dog's companions as he would be about those of his own children.

It will be readily understood that in the great sheep-feeding districts of Scotland there is no doggish crime so unpardonable as sheep-killing. As long as a dog can be kept from strange companions there is no great danger, as a collie is scarcely able to master the active and powerful sheep of those parts—sheep which, by reason of their semi-wild life, are able to defend themselves against foes to which a southern fold-bred sheep would at once succumb. But evil communications corrupt the manners of dogs as well as of men, and there is the greatest danger of several collies uniting in their attacks upon the sheep.

Some time ago a couple of shepherds met in a market place, each, as a matter of course, accompanied by his dog, one of which had been suspected of sheep-worrying. After the manner of dogs, the animals accosted each other, and soon assumed so remarkable a demeanor in their conversation that their owners consulted together on their own account, and agreed to set a watch upon their dogs. On that very evening both dogs started from their homes at the same hour, joined each other, and set off after the sheep.

Here we have a direct example that dogs have a sufficiency of language to convey ideas. The old offender had invited the young and innocent dog to go with him sheep-worrying, and had even managed to tell him the time when he was to start on his expedition. I have not been able to ascertain whether audible sounds were employed by the dogs, but I believe that the language, although perfectly understood by themselves and partly so by their masters, was entirely one of look and gesture.

An event occurred near Leslie which corroborates the story just told respecting dogs and their power of understanding their own language.

A farmer had lost a considerable number of sheep, and so he and his shepherd watched carefully throughout the night for the purpose of detecting the dog which had worried the animals. About the middle of the night they saw a troop of seven dogs making at full speed for the field where the sheep were kept. One dog was evidently the leader, and there could be no doubt that the animals, which belonged to different owners, had pre-arranged their meeting, and even settled the time at which they were to leave their respective homes. This could only have been done by means of some kind of language, which, though it did not consist of words, was as intelligible to them as human language is to mankind.

Two very remarkable instances of language and combination are given by Colonel W. Campbell in his "Indian Journal." The writer is perhaps better known by his *nom de plume*, "The Old Forest Ranger." He was at Ranee Bennore on a hunting expedition:

"I witnessed this morning a curious instance of wolfish generalship that interested me much, and which, in my humble opinion, goes far to prove that animals are endowed to a certain extent with reasoning faculties, and have means of communicating their ideas to each other.

"I was as usual scanning the horizon with my telescope at daybreak to see if any game was in sight. I had discovered a small herd of antelopes feeding in a field from which the crop had lately been removed, and was about to take the glass from my eye for the purpose of reconnoitring the ground, when, in a remote quarter of the field, concealed from the antelopes by a few intervening bushes, I faintly discerned in the gray twilight a pack of six wolves, seated on their hind quarters like dogs, and apparently in deep consultation.

"It appeared evident that, like myself, they wanted venison, and had some design upon the antelopes; and, being anxious to witness the mode of proceeding adopted by these four-legged poachers, I determined to watch their motions. I accordingly dismounted, leaving my horse in charge of the sowar, and, creeping as near the scene of action as I could, without being discovered, concealed myself behind a bush.

"Having apparently decided on their plan of attack, the wolves separated, one remaining stationary, and the other five creeping cautiously around the edge of the field, like setters drawing in a shy covey of birds. In this manner they surrounded the unsuspecting herd, one wolf lying down at each corner of the field, and the fifth creeping silently toward the centre of it, where he concealed himself in a deep furrow.

"The sixth wolf, which had not yet moved, now started from his hiding-place and made a dash at the antelopes. The graceful creatures, confident in their matchless speed, tossed their heads as if in disdain, and started off in a series of flying bounds that soon left their pursuer far behind. But no sooner did they approach the edge of the field than one of the crouching wolves started up, turned them, and chased them in a contrary direction, while his panting accomplice lay down in his place to secure wind for a fresh burst. Again the bounding herd dashed across the plain, hoping to escape on the opposite side; but here they were once more headed off by one of the crafty savages, who took up the chase in his turn, and coursed them till relieved by a fresh hand from an opposite quarter. In this manner the persecuted animals were driven from side to side and from corner to corner, a fresh assailant heading them at every turn, till they appeared perfectly stupefied with fear, and, crowding together like frightened sheep, began to wheel around in diminishing circles.

"All this time the wolf which lay concealed in the furrow near the centre of the field had never moved, and although the antelopes had passed and repassed within a few feet of him, and had, perhaps, even jumped over him, his time for action had not yet arrived. It now became evident that the unfortunate antelopes must soon be tired out; when it appeared probable that the surrounding wolves would have made a combined attack, and driven the terrified herd toward the centre of the field, where the wolf which had hitherto been lying in reserve would have sprung up in the midst of them, and secured at least one victim."

At this period of the proceeding the spectator shot the nearest wolf, whereupon the other five decamped and allowed the antelopes to escape.

Here we have reason and a power of combination for mutual action that would have done credit to human beings.

The anecdote shows also that there is much more detail in the language of animals than is generally supposed. Each had its different post assigned, so that the wolves must have possessed some means of indicating that locality; and each undertook to play its own part in a scheme of no small intricacy, so that their language must have been capable of expressing abstract ideas.

Mr. Walter Elliot, also a mighty Indian hunter, mentions in a foot-note to Colonel Campbell's account that he has witnessed similar instances of combination on the part of the same animal. Once he saw three gazelles chased by a single wolf. They made for a "nullah," or ravine, and plunged into it. Presently two of the gazelles bounded up the opposite bank of the nullah, but the third gazelle and the wolf were missing. Going to the nullah in order to discover what had become of the animals, Mr. Elliot found the missing gazelle in the jaws of *three* wolves. It was evident that it had been decoyed into an ambush, two wolves having hidden themselves in the nullah, and the third driven the gazelles to the spot where his accomplices were concealed, thus making up by cunning for lack of speed.

I rather think that each species has its own dialect, and that there is another language which is common to all—a sort of animal *lingua franca*, or "pigeon-English." For example, a cry of warning, no matter from what bird or animal it comes, is understood by them all, as is well known to many a sportsman who has lost his only chance of a shot by reason of an impertinent jay, crow, or magpie which has spied him, and has given its cry of alarm.

In Mansfield Parkyn's work on Abyssinia is a remarkable account of language and the consequent power of combination among the monkey tribe:

"You may see them quarreling, making love, mothers taking care of their children, combing their hair, nursing and suckling them; and the passions—jealousy, anger, love—as fully and distinctly marked as in men. They have a language as distinct to them as ours is; and their women are as noisy and fond of disputation as any fishfag in Billingsgate.

"The monkeys, especially the Cynocephali, who are astonishingly clever fellows, have their chiefs, whom they obey implicitly, and a regular system of tactics in war, pillaging expeditions, robbing corn-fields, etc.

"These monkey forays are managed with the utmost regularity and precaution. A tribe, coming down to feed from their village on the mountain (usually a cleft in the face of some cliff), brings with it all its members, male and female, old and young. Some, the elders of the tribe, distinguishable by the quantity of mane which covers their shoulders like a lion's, take the lead, passing cautiously over each precipice before they descend, and climbing to the top of every rock or stone which may afford them a better view of the road before them.

"Others have their posts as scouts on the flanks or rear; and all fulfill their duties with the utmost vigilance, calling out at times, apparently to keep order among the motley pack which forms the main body, or to give notice of the approach of any real or imagined danger. Their tones of voice on those occasions are so distinctly varied that a person much accustomed to watch their movements will at length fancy—and, perhaps, with some truth—that he can understand their signals.

"The main body is composed of females, inexperienced males, and young people of the tribe. Those of the females who have small children carry them on their back. Unlike the dignified march of the leaders, the rabble go along in a most disorderly manner, trotting on and chattering, without taking the least heed of any thing, apparently confiding in the vigilance of their scouts.

"Here a few of the youths linger behind to pick the berries off some tree, but not long, for the rear guard coming up forces them to regain their places. There a matron pauses for a moment to suckle her offspring, and, not to lose time, dresses its hair while it is taking its meal. Another younger lady, probably excited by jealousy or by some sneering look or word, pulls an ugly mouth at her neighbor, and then, uttering a shrill squeal highly expressive of rage, vindictively snatches at her rival's leg or tail with her hand, and gives her, perhaps, a bite in the hind quarters. This provokes a retort, and a most unladylike quarrel ensues, till a loud bark of command from one of the chiefs calls them to order. A single cry of alarm makes them all halt and remain on the *qui vive*, till another bark in a different tone reassures them, and they then proceed on their march.

"Arrived at the corn-fields, the scouts take their positions on the eminences all around, while the remainder of the tribe collect provisions with the utmost expedition, filling their cheek-pouches as full as they can hold, and then tucking the heads of corn under their armpits. Now, unless there be a partition of the collected spoil, how do the scouts feed? I have watched them several times, and never observed them to quit for a moment their post of duty, until it was time for the tribe to return, or till some indication of danger induced them to take to flight."

Here we have clear proof of the existence of a definite language among beasts — a language so expressive that it could be understood by a human listener. There are many birds which act in almost exactly the same manner, a few being posted as sentinels, while the rest devour the crops in peace, knowing that warning will be given if danger should threaten them.

The animal above mentioned is the Dogfaced Baboon. Colonel Drayson, R. A., has given a similar account of another species, the Chacma, of Southern Africa.

A ludicrous example of the possession of language of the monkey tribe is given by Sir J. Bowring in his admirable work on Siam. During a journey one of his suite fired at a monkey, wishing to secure the young one which she held in her arms. He did not kill her, and the wounded mother retreated into the jungle, carrying her child with her. The rest must be told in Sir John's own words:

"Five men immediately followed her; but ere they had been out of sight five minutes, we saw them hurrying toward us, shouting '*Ling, ling, ling, ling!*' (*i. e.*, monkey). As I could see nothing, I asked Mr. Hunter if they were after the monkeys.

"'Oh, no,' he replied; 'the monkeys are after *them*.'

"And so they were, thousands upon thousands of them coming down in the most unpleasant manner. As the tide was out, there was a great quantity of soft mud to cross before they could gain the boat. Here the monkeys gained very rapidly upon the men; and when at length the boat was reached, their savage pursuers were not twenty yards behind them.

"The whole scene was ludicrous in the extreme, and I really think that, if my life had depended upon it, I could not have fired a shot. To see the men making the most strenuous exertions to get through the deep mud, breathless with their run and fright combined, and the army of little wretches drawn up in line within twenty yards of us, screaming and making use of the most diabolical language, if we could only have understood them. Besides, there was the feeling that they had the right side of the question.

"One of the *refugees*, however, did not appear to take my view of the case. Smarting under the disgrace and the bamboos against which he ran in his retreat, he seized my gun and fired both barrels on the exulting foe, who immediately retired in great disorder, leaving four dead upon the field. Many were the quarrels that arose from this affair among the men."

This incident shows clearly the existence of language among the monkeys. Otherwise they could not have understood that one of their number had been injured by the hands of certain men, and so quickly have organized a combined attack upon their foes.

The following anecdotes have been sent to me by a London physician, and forcibly illustrate the faculty possessed by animals of communicating ideas to each other. The first is an example of dog language.

"While I was living in the country with a friend, a most interesting incident was observed in the history of the dog.

"My friend had several dogs, two of which had a special attachment to, and an understanding with, each other. The one was a Scotch terrier, gentle and ready to fraternize with all honest comers. The other was as large as a mastiff, and looked like a compound between the mastiff and the large rough stag-hound. He was fierce, and required some acquaintance before you knew what faithfulness and kindness lay beneath his rough and savage-looking exterior. The one was gay and lively, the other stern and thoughtful.

"These two dogs were often observed to go to a certain point together, when the small one remained behind at a corner of a large field, while the mastiff took a round by the side of the field, which ran up hill for nearly a mile, and led to a wood on the left. Game abounded in those districts, and the object of the dogs' arrangement was soon seen. The terrier would start a hare, and chase it up the hill toward the large wood at the summit, where they arrived somewhat tired. At this point the large dog, which was fresh and had rested after his walk, darted after the animal, which he usually captured. They then ate the hare between them, and returned home. This course had been systematically carried on for some time before it was fully understood."

The next anecdote shows that animals belonging to different species, such as the dog and cat, can communicate ideas to each other, and act in concert.

"A relation of mine in Dumfriesshire had a dog and a cat which were attached to each other in an extraordinary manner, and both were great favorites in the household. The dog, however, was not intended to sleep in the house, and was carefully put out every night; but, strange to say, he was always found in the morning lying before the fire, with the cat by his side.

"One evening the master of the dog heard a sort of rap at a back-door leading to the kitchen, and saw the sagacious cat spring up and strike the latch, while the dog pushed open the door and entered in triumph. This system must have long been carried on, and when it was discovered, I need not say how interested were the members of the household in these intelligent and really wonderful creatures."

Most persons have heard of the celebrated rook parliaments, though very few have seen them. I have an account written by a lady, who was at the time in bad health, and was reclining among some shawls behind a window-curtain, where even the sharp-eyed rooks did not detect her.

The account much resembles those that have already been given by other writers, but introduces one additional circumstance. The rooks (called crows by the spectator) assembled in a circle, and in the middle was one bird looking very downcast and wretched. Two more rooks took their places at its side, and then a vast amount of chattering went on. At last the two birds, which seemed to act as accusers, pecked the central bird and flew off. All the others then set on the condemned bird, pecked it nearly to pieces, and went away, leaving the mangled body on the ground.

The lady who witnessed this remarkable scene was much struck by the variety of tones employed by the birds, and their great expressiveness.

This account is corroborated by Major Norgate in his "Notes on the Indian Crow," published in the *Zoologist*, p. 9650:

"The crow has meetings for some reason or other; these the natives call Punchayeti—a sort of court.

"I have several times seen these assemblies. Four or five crows will alight upon an open space, generally on green grass. Two or three will begin cawing, and in a minute or two some forty or fifty of them will come flying toward the place by twos and threes from every quarter. They then form a kind of ring around one crow, which appears to have been an offender against some of the rules of their society, and they remain still for some minutes, the culprit never appearing to attempt to escape. Then, all of a sudden, five or six of them will attack the prisoner, pecking him, and striking him with their wings.

"On one occasion I saw the crow left dead on the spot, and on another the prisoner's wing was broken; but these courts, or whatever they are, suddenly come to a termination by the too near approach of a man or a dog. I saw one meeting which lasted twenty minutes; but no punishment was inflicted on any of them, and no noise was made. The whole assembly flew off together: they were not disturbed at all, and they were eating nothing, for it took place on a bare plain. Of course, it must only be surmised as to why these crows are punished by the others; perhaps some close observer may discover the reason."

Here is casually noticed a rather important fact, namely, that these crow parliaments are sufficiently common in India to have received a name in the language of that country, and that one individual saw several of them. I mention this, because several accounts of crow parliaments seen in this country have been received with considerable incredulity. The reader will observe that in all essential points the two narratives agree. My own correspondent is of opinion that the two birds which guarded the culprit were the accusers, and that it was their duty to inflict the first blow. There is a curious parallel here with that portion of the Mosaic law which ordained that in cases of capital punishment there must be at least two witnesses, and that they must cast the first stone at the convicted criminal.

An account of a similar act of justice is related by Mr. J. Drew, in Hardwicke's *Science Gossip* for October, 1871. The event occurred at Nansladron, in Cornwall.

"One summer afternoon my attention was drawn to a vast assemblage of rooks on our lawn. By the terrible vociferations they were making, it was evident that something very unusual was being enacted; for, clamorous as these birds are by nature, the noise and excitement of this meeting it would be almost impossible to describe.

"After watching them for some time, it became clear that they were in the act of carrying out some preconcerted punishment upon a luckless offender of their own flock; for on the ground was a black object in the form of a rook, which was evidently being pecked at, rolled over and over, and so passed on from rank to rank of the assembled multitude. That it was not a mere pastime was evident from the ruthless way in which feathers were pulled out and continuous blows given.

"Having waited about ten minutes, we felt a curiosity to know the effect of such chattering ferocity upon the poor black object, and drew near to pick it up. Of course the rooks flew away with loud cawings as soon as we approached; but, to our great astonishment, the prostrate bird opened its eyes, spread its ragged wings, and made, as it best could, for the nearest tree. Whether, if we had not interfered, the punishment would have been carried out *usque ad mortem* I know not. But clearly it was a good case to prove that the lower animals are governed by the same principles of thought and action as we are, each grade varying only in its mental and moral qualities in proportion to the development of the nervous system."

Here, as it will be noticed, the observer saw the infliction of the punishment, but not the trial which had evidently preceded it. Still he saw enough to show that the birds must have possessed the power of reasoning, and a language sufficiently definite to enable them to unite in a common object.

Other birds besides crows and rooks can assemble, hold council, and agree to act on the result of their deliberations.

One of my friends, then living near Manchester, in the garden had a very fine mountain-ash tree, which always produced a plentiful crop of berries. Shortly before the fruit ripened a great number of thrushes got together at the end of the garden, and were very noisy, chattering, and evidently discussing some subject on which they were not agreed. This went on for some time, the assemblage and chattering continuing daily. All this time the berries were ripening; and one morning an order appeared to be issued; the birds flew to the tree, and in a couple of hours there was not a berry left upon it. This occurred regularly during the three years in which my friend occupied the house.

Last year a somewhat similar event took place in the garden of one of my neighbors, who is a great horticulturist, and very successful with fruit as well as with flowers. There was a cherry-tree bearing in that year a remarkably heavy crop of fruit, which was carefully watched until it ripened. One evening the owner of the garden, seeing that the cherries had just reached the proper stage for picking, ordered the gardener to gather them on the following morning. But the birds seemed to know as much about fruit as he did, for when the gardener came with his basket the crop of cherries had vanished, and nothing was left except the stalks, each with the stone still attached to it.

It was evident that in this case the birds must have entered into some agreement on the subject, and must have arranged among themselves not to meddle with the tree until the fruit was quite ripe. The disappointed owner of the cherry-tree stoutly avers that the birds overheard him give the order to the gardener, and so anticipated him; but the former anecdote, showing the power of mutual arrangement among birds, explains the latter.

An example of a somewhat similar mode of action was related by Mr. G. B. Clarke, of Woburn, to the Rev. F. O. Morris, and by him published in the *Naturalist*:

"In the summer of 1849 a pair of martins built their nest in an archway at the stables of Woburn Abbey, Bedfordshire; and as soon as they had completed building it, and had lined it, a sparrow took possession of it, and although the martins tried several times to eject him, they were unsuccessful. But they, nothing daunted, flew off to scour the neighborhood for help, and returned in a short space of time with thirty or forty martins, who dragged the unfortunate culprit out, took him to the grass-plot opposite, called 'The Circle,' and there fell on him and killed him."

This story was told by Mr. Clarke to Mr. Morris a few days after its occurrence. It is useful in this place as showing that birds are able to communicate their thoughts to each other by means of a language. Supposing that we had heard the aggrieved martins talking to their friends, we should have distinguished nothing but a meaningless twitter. But, even with human beings, especially those who are uneducated, the sound of a strange language is scarcely more intelligible than the twittering of birds or the bleating of sheep; and, indeed, the well-known term of Barbarian—*i. e.*, those whose language is nothing but "bar-bar"—shows how the sound of an unknown language affected even the well-educated and cultivated Athenians.

It is not likely that in the language of animals there are any principles of construction such as are possessed by all human languages. But the same effect may be produced by different means, and the reader will see that in this instance no human language, however perfect its construction, could have served its purpose better than did the inarticulate language of the birds. They told their friends that their dwelling was usurped by an intruder too strong to be ejected by them; they asked for united assistance, and arranged the course to be pursued. Had not this been done, it is evident that the birds could not have acted so perfectly in concert.

In fact, wherever animals of any kind form alliances and act simultaneously for one common object, it is evident that language of some sort must be employed.

Here is a case where one dog saw another in difficulty, and went to give it advice. Finding that its advice was not taken, it went again, and forced the reluctant animal into action.

The dog, a little black-and-tan terrier named "Beau," and his owner were at Penmaenmawr, on the coast of North Wales. They were one day on the sands, and were overtaken by the tide, which cut them off from the shore by a belt of water. A bathing-machine came up and took off the dog's owner, Beau refusing to enter the machine, of which he seemed to be suspicious. The rest must be told in the writer's own words, taken from the account in *Old and New*, for December, 1873:

"When I found myself on the beach, I looked for my dog, thinking that he would probably come swimming after the machine. But no; the little idiot was still on the island, yelping and barking in great distress. I called to him for a long time, bidding him swim across, as I knew that he could use his limbs almost as well in water as on land. But the naughty animal would not come, and meanwhile the sea was gaining on the sand, and Beau had scarcely space to stand and whine.

"Playing near me on the beach was a large, rough-haired, brave dog—a sort of half-bred retriever, I should suppose. He perceived the fix we were in, and suddenly dashed through the water and went up to Beau, and said something to him. I don't know what he said; but I have no doubt that he counseled Beau to swim across to his mistress. Alas! the kind, brave dog returned to dry land, but no Beau. By this time the sea had risen round my little terrier, and he was himself like a tiny black-and-tan island.

"Now what did the brave dog do? For the second time he dashed through the water and stood beside the shivering, yelping creature; then he went behind Beau, and very gently but firmly pushed, pushed, pushed him through the water toward the place where I was standing. As soon as they were both fairly in the deep sea, and it seemed to be a case of sink or swim with Master Beau, the wise, brave dog let him go, and with a few vigorous strokes brought himself to shore. Beau, having received such an impetus, very soon presented himself dripping and breathless at my feet, amid the applause of the assembled multitude. The brown dog, like a true hero, made no fuss about what he had done, and I had nothing to give him but a pat on the head. His master was certainly not on the beach at the time, and I do not think I ever saw the dog again."

In the well-known *Science Gossip* there is a very interesting paper by Mr. E. F. Elwin on the habits of an ant called *Myrmica ruginodis*. As is their manner, two of them had been fighting, and one had succeeded in catching his opponent by one of the antennæ. Ants always try to do this, as, if they succeed, the adversary succumbs at once. In fact, with regard to ant combatants, the result of seizing the antennæ is precisely that which is known among pugilists as "getting the head into Chancery," namely, rendering the opponent helpless.

Another ant, coming up, seized the victim by a leg, and tried to pull it away, but in vain, and though a crowd assembled round the combatants, they could not put an end to the fight. At last a single ant ran up and stroked with his antennæ the victor, who at once released the prisoner, and both the combatants and the spectators went quietly away.

This is another example of an animal assisting its fellow-creature, and doing so by means of its own language, when force had proved unavailing.

The following remarkable instance of the communication of ideas among the lower animals is narrated by the Rev. C. Otway:

"At the flour-mills of Tubberakeena, near Clonmel, while in the possession of the late Mr. Newbold, there was a goose, which by some accident was left solitary, without mate or offspring, gander or goslings. Now it happened, as is common, that the miller's wife had set a number of duck eggs under a hen, which in due time were incubated; and, of course, the ducklings, as soon as they came forth, ran with natural instinct to the water, and the hen was in a sad pucker—her maternity urging her to follow the brood, and her instinct disposing her to keep on dry land.

"In the mean while up sailed the goose, and with a noisy gabble, which certainly (being interpreted) meant, 'Leave them to my care,' she swam up and down with the ducklings, and when they were tired with their aquatic excursion she consigned them to the care of the hen.

"The next morning, down came again the ducklings to the pond, and there was the goose waiting for them, and there stood the hen in her great flusteration. On this occasion we are not at all sure that the goose invited the hen, observing her maternal trouble; but it is a fact that she being near the shore, the hen jumped on her back, and there sat, the ducklings swimming, and the goose and hen after them, up and down the pond.

"This was not a solitary event; day after day the hen was seen on board the goose, attending the ducklings up and down, in perfect contentedness and good-humor—numbers of people coming to witness the circumstance, which continued until the ducklings, coming to days of discretion, required no longer the joint guardianship of the goose and the hen."

D

On the evening of January 15, 1874, I received a remarkable corroboration of the truth of this story. I was narrating it to a lady, who I found was perfectly acquainted with the facts. She had heard the story told by a friend of hers, who had witnessed the curious alliance between the hen and the goose, and had not the least idea that it had ever appeared in print.

There are one or two points about this narrative which are deserving of notice. That language was employed by the goose, the hen, and the ducklings, is evident enough; but it is a curious question whether the ducklings understood the hen better than the goose, or *vice versa*. I am rather inclined to think that when a hen tries to call from the water the ducklings which she has hatched, she fails because she does not know how to express herself. Her own chickens would never venture into the water, and she has no words in her vocabulary to suit the occasion.

Ducklings understand a duck well enough; but when they are in the water they do not pay the least attention to the hen on the land, though she may flutter about in the greatest distress, and use every means in her power to call her foster-children to the shore. It seems, in this case, as if the aquatic goose could talk to the aquatic ducklings, both having the same expressions in their vocabularies. It could take charge of them as long as it thought proper, and, when the time came, order them ashore, and deliver them over to the hen. They did not obey, or did not understand the hen, when she called them to come on shore; but they both understood and obeyed the goose.

That there was also a language common to both parties is evident from the action adopted by the hen. She could not have sat on the back of the goose unless invited by the latter, which, as we shall see in the course of the work, is a bird possessed of great intellectual powers.

CHAPTER VI.

LANGUAGE—[HUMAN].

Necessity for Communication of Ideas between Man and the Lower Animals.—The Latter able to Make Themselves Intelligible to Man.—The Gander and the Goslings.—The Skye Terrier and the Distressed Kitten.—Gesture-Language of Cats.—Language of Intonation in Man and Animals.—Gesture-Language Employed by Animals as well as Man.—Gesture-Language of the Rat.—Capability of the Animals to Understand Human Language, even when not addressed to them.—The Dray-horse and their Drivers.—"Turk," the French Dog.—A Parrot Speaking Two Languages.—Various Parrot Stories.—The Mastiff Overhearing the Midnight Conspiracy.—The Retriever Understanding his Master and Anticipating Him.—"Rory" and "Banquo" Obeying various Orders.—How to Teach Animals.—"Ned" and the Rabbit.—"Carina's" Pitiful End.—A Canine Umpire between a Farmer and his Shepherd.—A Canine Connoisseur in Wools.—"Sweep" and the Cows.—Baldie Tait's Collie Dog "Hastie" and his Dog "Susy."—How the Collie Dog "Watch" Understood his Master and Helped Him out of a Scrape.—"Ben" Evading an Overheard Order of his Master.—"Help" Overhearing and Evading an Order for his Execution.—Another Dog Acting in a precisely Similar Manner.—Dodge and Counter-dodge.—"Bijou," the Spitz Dog, Accepting a Reproof and Altering his Behavior.—The Hon. Grantley Berkeley's Dogs, and their Comprehension of Human Language.—"Missy" Understands the Doctor's Order, and Acts upon it.—The Cat "Rosy" Sent upon a Message to a Lady, and Delivering it Intelligibly.

THE next branch of the subject extends to man as well as beast. We have seen that the beasts possess a language by which they can communicate ideas to each other, and that they can act upon the ideas so conveyed. We have now to see whether they can convey their ideas to man, and so bridge over the gulf between the higher and the lower beings. Indeed, if there were no means of communicating ideas between man and animals, domestication would be impossible.

Every one who has possessed and cared for pet animals must have observed that they can do so. In many cases even their own language becomes intelligible to man. Just as a child that can not pronounce words expresses its meaning by intonation, so there is all the difference in the world between the different modes of barking of the same dog. There is the bark of joy or welcome, when the animal sees its master, or anticipates a walk with him. Then there is the furious bark of anger, if the dog thinks that any one is likely to injure himself or his master. And there is the bark of terror, when the dog is suddenly frightened at something which it can not understand. Supposing that its master could not see the dog, but only heard it bark, he would know perfectly well the ideas which were passing through the animal's mind.

Take the cat. Every one who has kept cats knows the difference between the mew of distress and that of ordinary conversation, the purr of pleasure—the little gratified chuckle of content when touched by friendly hands, and the low, loving tones in which a mother-cat talks to her kittens. It is the same with birds. A pet canary, for example, always knows how to call its master, and when it sees him will give a glad chirrup of recognition quite distinct from its ordinary call.

The bees and wasps have quite a different sound in their wings when they are angry to that which they emit when only at their ordinary work. It is a distinct menace or challenge to a supposed enemy, and any one who is conversant with the ways of these insects understands it, and makes the best of his way off.

These are examples of sound-language, while the gesture-language is wonderfully extensive and expressive. If a cat were to say in so many words, "Please open the door for me," it could not convey its ideas more intelligently than it does by going to the door, giving a plaintive mew to show that it wants help, and then patting the door. The dog, or, indeed, any animal accustomed to live in the house, will act after a similar fashion.

Here, then, we see that the lower animals can form connected ideas, and can convey them to man, so that the same ideas are passing at the same instant through the minds of man and beast, showing that they possess the same faculties, though of different extent.

The following anecdote of a rat shows how expressive and intelligible is the language of gesture: A gentleman living in Kent had a fancy for taming animals, and among others had some rats, which were on the most friendly terms, and used to run about him as he sat in his room. One of his rats had a litter of young, and, in order to insure their safety, they were placed in a bird-cage and hung on a wall.

One night, after their master was asleep, he was awakened by something patting his cheek, and found it was one of his tame rats. He tried to sleep again, but the animal would not allow it, and was evidently disturbed about something. As soon as he obtained a light, the rat went to the door, and looked at him as if it expected him to follow. He did so, and it led him down the stairs into his room, and took him to the spot where one of the young was lying, having fallen from its cage.

Suppose that we substitute for the rat a deaf-and-dumb man or woman, the action would have been almost exactly the same, as would have been the ideas that were so lucidly conveyed by the language of gesture. The animal found that it was unable to put back its fallen young one, and must have calculated that its master was taller and stronger than itself, and able to replace the young rat. So it went in search of its master, traced him to his bedroom, which it must have done by the sense of smell, awoke him from his sleep, and showed him where his assistance was needed.

An anecdote of a somewhat similar character will be related under another heading, the actor being a dog instead of a rat. Indeed, most of these anecdotes illustrate several characteristics common to man and beast. I might have placed this present anecdote under the heading of Parental Love or Reasoning; but as it shows that, by means of their own language, beasts can convey their thoughts to man, I have placed it in its present position.

How completely animals can make themselves understood by man, especially when they wish to help each other by the aid of man, will be seen in many of the anecdotes narrated in this work. Here is a case where a gander managed to convey ideas to human beings:

"I was once sitting at my window reading, when a gander came up and stood at the window, uttering the most discordant screams, and making the strangest gestures with his head. I was aware that he was a knowing bird, but was not prepared for the sequel.

"As soon as my wife and I came out he waddled away round the stables and out-houses until he came to the mill-wheel. Then he stopped, went forward a few paces, and kept looking round at us. We could see nothing wrong; but in a short time we heard the plaintive voice of some young goslings which had fallen through the mill-lade, which had been left open.

"There was no possibility of rescue except by putting on sufficient water to wash them through the conduit. I did so, ran to the end, caught them as they were washed out, and restored them to their delighted parent. The gander seemed overjoyed, as could be seen by his action as he strutted off to a place of safety, conscious that he had done great things. So he had."

As an example of gesture-language, nothing could be more clear and intelligible than the method employed by a Skye terrier belonging to one of my correspondents.

He had formed a friendship with a kitten, and the two were one day in the garden. Presently the kitten wished to go into the house, and finding the door shut, tried to call the attention of the servants by mewing under the window. She could not succeed in making them hear, whereupon her friend, the Skye terrier, picked her up gently in his mouth, held her in front of the window and shook her backward and forward so as to be seen by the servants. They understood what the animal meant, let the kitten into the house, and ever afterward the dog employed the same expedient. It is exactly that which would have occurred to a human being under similar circumstances.

On account of the exigencies of space, I am obliged to omit many anecdotes which show the power of gesture-language in the lower animals. I must, however, mention one or two more. I have at the present time a cat which is not as companionable as I like to see a cat, being rather of a retiring and self-seeking disposition.

Nevertheless, she is quite aware of the fact that I can understand her language, and always comes to me in any difficulty. She is rather given to straying, I fear, in some poaching raids upon a neighboring rabbit-warren, and consequently finds herself locked out of the house. When this is the case, she jumps on the sill of the window, raises herself on her hind legs, so as to peer above the dead-glass blind, looks at me, and sets up a most piteous mew, or rather howl. No sooner do I rise than she jumps down, and before I can reach the door she is already there, purring and rubbing herself against it in anxious expectation.

She comes in very slowly, gives a passing greeting, and then goes off to the kitchen, where she has two children, who are quite as big as herself, and all three coil themselves up into an indefinite heap of black and white fur, in which a head, a tail, or a leg occasionally shows itself without any particular reference to any individual animal.

A correspondent has furnished me with a very similar account of her own cat, "Daisy." In almost exactly the same manner the cat used to make herself very conspicuous at the window. Her mistress would then point toward the door. The cat, having made her own gesture-language intelligible, understood that of her mistress, and went to the door in certain expectation that it would be opened for her.

Examples of animals making their language intelligible to man could be multiplied *ad infinitum*, and I therefore pass to the next division of the subject, namely, the capability possessed by the lower animals of understanding the language of man.

That many of the lower animals understand something of human language is a familiar fact. All the domesticated animals, especially the dog and the horse, can comprehend an order that is given to them, though, perhaps, they may not be able to understand the precise words which are used. Yet there are many occasions, it is evident, when the knowledge of human language does extend to the signification of particular words.

Some of my readers may remember the elephantine dray-horses which were engaged to draw the funeral car of the late Duke of Wellington. When the time for starting arrived there was a hitch in the proceedings, for the horses could not be induced to move. At last some one hit upon the reason, and brought a drayman, who said, "Gee, there!" or words to that effect, on which the animals started at once. These horses are never beaten, are always treated with kindness, and are directed entirely by voice, the long whip being only used for ornament, or for gently stroking the animals.

There is a French dog called "Turk" near my house, who was in a very uneasy state of mind for some time after he came to England. He did not know English, and was as puzzled as if he had been a human being under similar circumstances. If addressed in French, he seemed quite delighted and at his ease; but it was not for some time that he learned English sufficiently to be comfortable.

There was a parrot, well known to our family, which was able to speak in two languages, and, when addressed, always replied in the language used by her interlocutor, speaking English or Portuguese, as the case might be.

I never yet met with any owners of pet talking parrots who had not come to the conclusion that the birds not only imitate human language, but that they understand the signification of the words which they utter, and use them accordingly. I personally knew two parrots who, if the servant neglected to feed them at the proper hour of the morning, would call her by name, and shout loudly for breakfast. There was another parrot —a green one—whom I did not know in life, having only seen her preserved skin in a glass case. She was a great favorite with the family, being allowed to go at large over the house, and in consequence was brought into much closer relationship with human beings than is generally the case with birds.

None of the family had the slightest doubt that Polly was quite as well acquainted with the meaning of the words which she spoke as any of them could have been. Sometimes, before her feeding-time, she would call out, "Cook! cook! I want potato." She knew what potato was as well as the cook did, and if any thing else was put in the pan she would take the vessel in her beak, throw out all the contents, and then cry, "Won't have it! turn it out!"

Now she had never been taught either the deed or the words. When she arrived in the family she was new from her voyage, and could only speak a sort of jabber, called by the sailors "bush-talk," probably picked up from the natives, together with a very few expressions, most of which were of a nautical and decidedly objectionable character. In all probability she had noticed one of the servants use those words when throwing something away which she disliked, and had imitated her both in word and in gesture.

On another occasion one of the children, who was then about seven or eight years of age, had been reading about a mode of secret writing by means of lemon-juice, and was fired with a desire to try the experiment for herself. There did not happen to be a lemon in the house, and so she thought that she would try what vinegar would do. One of my children, by the way, took just the same idea a few months ago.

The only way to get at the vinegar was by intercepting the cruets as they were brought out from her parents' dinner. So she placed herself in readiness in the kitchen, took the vinegar, and was pouring it into a spoon, when she was interrupted by the parrot, who called out, "I'll tell mother! Turn it out! turn it out! turn it out!" Whereupon the conscience-stricken child threw away the cruet and the spoon, and ran off to the nursery as fast as she could. She had the fullest belief that the parrot really would tell her mother.

The few scraps of language which she had learned on board ship were occasionally produced just where they ought to have been omitted. On one occasion the remarks were so singularly inopportune that one of the family offered a remonstrance, saying, "Oh, Polly! Polly! who *could* have taught you such language?" Whereupon the bird at once replied, "You did." It is impossible, or, at all events, in the highest degree improbable that the bird should not have understood the language of its interlocutor as well as herself.

Being in a family almost entirely composed of girls, Polly had an objection to the opposite sex, especially in the form of boys. On one memorable occasion some boys had come on a visit, and, after the manner of their kind, became very uproarious. At last Polly could endure it no longer, but called to one of the daughters of the house, "Sarah! Sarah! here is a hullabaloo!" Parrots, by the way, have a curious predilection for the name of Sarah, which seems especially easy for them to pronounce.

The same parrot always looked out for the presence of the mistress of the house at the breakfast-table. If she did not come down before the meal was begun, Polly would begin to inquire after her in a plaintive tone, "Where's dear mother? is not dear mother well?" and so on, evidently having heard and understood similar comments by members of the household.

A very similar circumstance is related of a parrot by one of my correspondents.

It was an established custom in the household that at evening prayers the dog and the cat were to accompany the servants. One evening the dog made his appearance without his usual companion; whereupon the parrot called out, "Where's Cattie?" this being the familiar name by which the cat was called in the house.

Instinct is quite out of the question in any of these cases. The bird had first used its reasoning powers, and had then communicated the result to human beings in their own language.

The following anecdote, related by the late Rev. Cæsar Otway, who produces vouchers for the exact truth of the story, affords a remarkable instance of the capability possessed by the lower animals of understanding the language of man:

"A gentleman of property had a mastiff of great size, very watchful, and altogether a fine, intelligent animal. Though often let out to range about, he was in general chained up during the day.

"On a certain day when he was let out he was observed to attach himself particularly to his master. When the servant came, as usual, to fasten him up, he clung so determinedly to his master's feet, showed such anger when they attempted to force him away, and altogether was so peculiar in his manner, that the gentleman desired him to be left as he was.

"With him the dog continued the whole day; and when night came on, still he stayed; and on going toward his bedroom, the dog resolutely, and for the first time in his life, went up with him, and, rushing into the room, took refuge under the bed, whence neither blows nor caresses could draw him.

"In the midst of the night a man burst into the room, and, with dagger in hand, attempted to stab the sleeper. But the dog started at the robber's neck, fastened his fangs in him, and so kept him down that his master had time to call for assistance and secure the ruffian, who turned out to be the coachman. He afterward confessed that, seeing his master receive a large sum of money, he and the groom conspired together to rob and murder him, and that they plotted the whole scheme *leaning over the roof of the dog's kennel*."

The foregoing statement does not assert that the dog understood human language as completely as the men themselves did. But it is evident that the animal did gather from the conversation of the men that they intended to injure his master. The narrator does not state whether the conspirators mentioned any particular time for the murder, which was probably left to opportunity. The companionship of the dog during the day (which the intending murderers knew) might have prevented them from attacking their master by daylight, while his presence at night (which they did not know) effectually counteracted their plot.

Here is another anecdote, which shows that an animal is capable of understanding human language even although it be not addressed to it personally. A gentleman who possessed a very intelligent retriever dog was going from home for some time, and arranged that the dog should be sent to the house of a friend during his absence. On the day fixed for his departure the dog went on his own account to the house, and there remained until his master's return. When his master did come back the dog was overjoyed to see him, but became uneasy at the long call which was being made. He evidently took it into his head that his master was meditating another absence, and every time that he heard the hall door shut he rushed up-stairs to make sure that his master was in the house. At last, losing patience, he took his master's hat out of the hall, and carried it up-stairs to him, as a broad hint that he had better go home.

The following quaint anecdote is told by the late Charles Dickens, and is given in Forster's "Biography." It is evident from internal evidence that the district was given to brick-making. The story illustrates the capacity of the dog for understanding human language, and conveying ideas to human beings:

"I must close (14th of May, 1867) with an odd story of a Newfoundland dog—an immense, black, good-humored Newfoundland dog.

"He came from Oxford, and had lived all his life at a brewery. Instructions were given with him that, if he were let out every morning alone, he would immediately find out the river, regularly take a swim, and gravely come home again. This he did with the greatest punctuality, but after a little while was observed to smell of beer. She was so sure that he smelled of beer that she resolved to watch him.

"Accordingly he was seen to come back from his swim round the usual corner, and to go up a flight of steps into a beer-shop. Being instantly followed, the beer-shop keeper is seen to take down a pot (pewter pot), and is heard to say,

"'Well, old chap! Come for your beer, as usual, have you?'

"Upon which he draws a pint and puts it down, and the dog drinks it.

"Being required to explain how this comes to pass, the man says:

"'Yes, ma'am; I know he's your dog, ma'am; but I didn't when he first come. He looked in, ma'am, as a brick-maker might—and then he come in, as a brick-maker might—and he wagged his tail at the pots, and he give a sniff round, and conveyed to me as he was used to beer. So I drawed him a drop, and he drank it up. Next mornin' he come agen by the clock, and I drawed him a pint, and ever since he has took his pint regular.'"

My own dog "Rory" perfectly understood much of our conversation, and if told by any of us to fetch the slippers, to shut the door, to wipe his feet, or to put the cat down-stairs, he always performed the right act, showing that he knew the ideas represented by different words.

I know a dog, named "Banquo," who has learned to wipe his feet on the mat when he goes to a strange house; but on no consideration can he be induced to do so in his own house, where he considers himself privileged to do as he likes. Now my Rory acted in a very different manner; for he always wiped his feet whether they required it or not, and would never think of entering a room until he had rubbed all his feet for some little time.

In connection with this habit, I must mention the case of a gigantic Newfoundland dog belonging to a clergyman. He had not learned to wipe his feet, but he did know when they were dirty, and acted accordingly. When he came in with dirty feet, he crept into the hall gently, and so up-stairs, taking care not to allow his foot-steps to be heard. But when his feet were clean, he would clatter up and down stairs, making almost as much noise as a pony.

The dog "Banquo," who has already been mentioned, has a fine capacity for imitating a lady in hysterics. When told to go into hysterics, he sits in his mistress's lap, howls, yelps, flourishes his paws in a most fantastic manner, and ends by flinging himself backward. But he never performs this last feat without looking to see that a protecting arm is ready to catch him as he falls, thus adding to the truth of the representation. I have often seen him go through this performance, and a most ludicrous one it is.

The lady who possesses this animal, and who has taught it many comical tricks, sends the following remarks on the training of dogs:

"Any one wishing to educate dogs should commence by teaching them a few simple words (not blows), with occasional rewards for proficiency in any accomplishment. Twilight, and the dim but cheerful light of the winter fire, seems a fitting time for a pause in the day's work or the day's amusements, and then our little dog Banquo thinks himself entitled to share in the conversation. Last winter, having taught him that his two fore paws are his hands, I showed him how to warm them by sitting up and holding them outstretched to the fire. I remember at a friend's house seeing three cats on three footstools, in undisturbed possession of the dining-room fire. Our dogs are never allowed thus to monopolize either fire or hearth-rug, therefore the command 'Come and warm your hands' is generally most willingly obeyed."

The following anecdote is sent to me by a gentleman living at Bassendean:

"I had a Bedlington terrier, called 'Ned,' a very clever and intelligent dog. A few months ago I was in the Bassendean bog, when Ned started a rabbit among some whins. The rabbit ran toward the dike, and escaped its pursuer by getting into a hole in the dike. Two men happened to be passing at the time, and, though strangers to the dog, they helped him by pulling away some stones at the place where the rabbit had concealed itself.

"Ned immediately sprang into the hole, caught the rabbit, and, after killing it, ran off with it to me, a distance of three or four hundred yards. I said to the dog, 'Ned, you scoundrel, how dare you take the rabbit from those men?' The words were scarcely out of my mouth when my dog started off as fast as he could run, with the rabbit in his mouth, and laid it at the strangers' feet. "Any dog may be trained to carry things to his master, but this is the only instance I know of where the dog took the thing to a stranger. It certainly showed that he understood my words.

"When I am out in the fields, Ned is always told to take care of my coat, in the pocket of which I often have large sums of money. He has sometimes been left in the field alone until eight o'clock at night, and, although the field has been full of laborers and Irish reapers, not one of them has dared to touch the coat. The dog is peculiarly mild and gentle in his temper, but he will not permit any one to touch his master's property."

The following pathetic little story is from the same source:

"Mr. H—— had a beautiful little Blenheim spaniel, called 'Carina.' About the beginning of 1873, while the family were from home, the gardener slept in the house to take care of it.

"One night Carina, who had a family of healthy puppies about a fortnight old, came to the man's room, and scraped at the bedclothes until he awoke. Without striking a light or examining the dog in any way, the man said, 'Carina, go back to your puppies,' and the dog accordingly went away. In a short time she came again, and awoke the man in the same way. She again received the same order, and obeyed it as before. In the morning, when the gardener went to look at the dogs, the puppies were quite well, and Carina was lying by their side quite dead. Her puppies survived, and were brought up on cow's milk."

It is evident that the poor little dog felt her end approaching, and tried to make her last farewell before she died. That she was not understood was not the fault of the dog, but of the man, who was too dull or too sleepy to comprehend her meaning, though she could understand him.

Here is an account of a dog which shows that animals who live much with mankind manage to learn more of human language than is generally supposed. It was sent to me by a lady who knew the dog, the collie named "Moss," who has already been mentioned:

"His master and the shepherd were employed in moving sheep from one part of the farm to another. On reaching a certain point they fell into a dispute about the number of the sheep, the shepherd saying that they had the proper number, while the farmer thought that there ought to be one more. Not being able to decide, they jokingly appealed to Moss. The dog at once started off, and presently returned, driving before him the missing sheep, which he had brought from a spot quite out of sight, and a considerable distance on the opposite side of a hill."

Collie dogs are noted for the manner in which they can understand their masters' wishes; and the following anecdotes sent me by a Scotch gentleman show that they not only comprehend his general meaning, but the actual signification of his words:

"For several years within the last half century a deceased friend of mine was extensively engaged in the wool trade, and was a considerable buyer in Dumfriesshire. In one of these journeys, and after a forty-mile drive in his gig, he reached the house of a hill farmer in that country, arriving just at the close of the day. The farmer told him that his samples of wool were at some distance from the house, and that he would submit them for inspection on the following morning. My friend met with an hospitable reception, and as the hours of evening glided on the conversation turned on the management of sheep and cattle, and especially on a fine breed of shepherd dogs possessed by the farmer.

"Early in the morning all were astir, and the farmer and his visitor left the house for the purpose of examining the different kinds of wool. But great was the astonishment of my friend when they reached a level patch of ground between high hills where there was nothing to be seen but a shepherd and two dogs, to be told that this was the place for inspecting the wool.

"He was asked which kind of wool he would look at first, and, having named the kind, the shepherd called one of the dogs, and directed him to turn the sheep upon one of the hills, and bring them to him. The wise animal bounded off, and in a very short time the sheep were seen descending the hill by an easy pathway. The wool was examined, and the sheep driven back to the hill by another road. In the mean while the second dog was sent to bring forward another breed of sheep from a different place, and so on until all was finished, and without the least confusion. This plan was followed by the worthy farmer because he had not been able to find time to clip his sheep."

J—— P——, an elder in a Border congregation of the United Presbyterian Church, and a very truthful and worthy man, lately informed me that when he was a young lad he was at service at B—— farm, in Berwickshire, and had charge of the cattle. In the discharge of his duties he was accompanied by a very intelligent collie dog, called "Watch."

The farm is bounded on the west by the river Whiteadder. Its stream is comparatively trifling in dry weather; but, owing to the drainage of the high lands on its banks, whenever rain falls in any quantity hundreds of little rills pour into the channel of the river, so that in a very short time it overflows its banks. For the same reason it diminishes rapidly when the rain ceases.

On one occasion, the day being stormy and cold, he went into one of the cottages to warm himself, and on coming out he observed that one of the "kyloes" had strayed from the rest. On looking about, he saw the missing animal grazing among some cattle belonging to another farm on the opposite side of the river. During his absence in the cottage a rain-storm had come on, and the river had risen to a flood, so that he found it impossible to cross it and bring back the strayed beast.

Not knowing what to do, and without any expectation that he would be understood, he said to the dog, "Watch, I canna gang through to fetch the kyloe; ye'll hae to gang." The intelligent animal immediately plunged into the rapid rolling water, and reached the opposite bank. He went straight to the animal which belonged to his master, paying no regard to the others which were grazing with it, and brought the beast safely across, both animals being obliged to swim.

By so doing he helped his master out of a scrape; for the kyloe could not have strayed if he had not neglected his duty by staying in the hut long enough to allow the river to rise.

The same man, when engaged on another farm, had a collie dog to help him. One day, after the cattle had been driven into the sheds, he found that he must remain for some time longer in order to fodder them. He turned to the dog and said, "I dinna need ye any mair to-night, an' ye had better gang hame noo." The dog perfectly understood him, and went home at once.

"'Ben,' a very fine collie, belongs to an acquaintance of mine, a farmer. One day, as Ben's master was preparing to go to a village at some miles' distance, his wife asked whether he meant to take Ben with him. He answered that he should not do so, and told her to lock up the dog until he came back. Ben, hearing this, slipped out of the house unperceived; and when his master reached the village, he found Ben waiting for him."

It is evident that in this case the dog must not only have understood that he was not to go, and that he would be locked up in order to keep him at home, but that he must have known and recognized the name of the village which his master was about to visit.

"On one occasion when the farmer was going to Berwick by train from the village which has just been mentioned, the faithful Ben had contrived to follow him, and sprang into the carriage just as the train moved off, so that he could not be turned out. The dog attended him all day until his master was about to leave. Time was up, so that he could not wait for the dog, but went off in the train to the station whence he had started, and thence to his home. He had only reached his house for a very short time, when Ben presented himself, all covered with mud, and quite flustered with fatigue, having evidently run the whole distance, some thirteen miles, at full speed."

Reference will be made to this branch of the subject under the title of "Love of Master."

A Scotch gentleman has kindly forwarded to me the two accompanying stories, which illustrate the wonderful capacity enjoyed by many dogs of understanding even the minutest of their masters' language:

"A son-in-law of mine, an extensive sheep-farmer in Berwickshire, Mr. G——, of C——, had a collie dog, 'Sweep,' one of the very best of his kind. When, on account of old age, he became unfit for his ordinary work, he was used for taking out and bringing in the cattle from the parks. He generally lay before the kitchen fire, and, when milking-time came, all that was required was to say, 'Sweep, go for the cows,' when he would at once get up and go for them, bringing them up to the byre without any assistance whatever.

"It sometimes happened that he would leave a cow behind in the field; but whenever he was told so, he would again start off, pick out the cow from among the young cattle, and take her to the byre with the rest. I regret to say that poor Sweep is dead. Without any premonitory symptoms of illness, he was found lying cold and stiff one morning in front of the shepherd's house.

"Some time about the beginning of the century there lived on Clint's farm a man of the name of 'Baldie Tait,' a noted sheep-stealer. He had a collie as accomplished a thief as himself, and there are those still alive who have known him to direct this dog to go to Heriot Muir, a distance of several miles, to pick out the best sheep he could get, take them to Hangingshaw, a wayside public-house on the high-road to Edinburgh, and remain with them till Baldie should come, when they were driven to Edinburgh and sold.

"Baldie had become so notorious in his unlawful calling that a warrant was issued for his apprehension. On the day on which he heard of this he went to a sale of farm stock at a place called Muircleuch, near Lauder. He as well as his dog were well known; and Baldie, knowing how to improve the occasion, put up the dog to be sold by auction, got ten pounds for it, and decamped at once—not a minute too soon, for in a short time the minions of the law were on his track, but they were too late."

On receiving this account, I wrote to the narrator, expressing my surprise that even so great a rascal as Baldie should have parted with his dog. I found, however, that selling the dog was a way he had when he wanted money; for, by some means unknown, he always got the animal back again.

One of my correspondents has favored me with a brief history of a dog which understood and obeyed the orders of its master:

"Not many years ago there lived in Edinburgh a drunken, shiftless mole-catcher, of the name of Hastie. Like most of his trade, he was very fond of dogs, especially terriers; and he had one which he loved above all others, calling her his 'blessed Susie.'

"She often used to act the part popularly attributed to the jackal, and provide her master with food. When, as often happened, Hastie was in straits for food, and had no money, he would go past a butcher's-shop, point to a piece of meat, and say quietly, 'Susie, I want that.' He then went on his way, and in a few minutes the meat was sure to be in his possession.

"Time wore on, and, as every thing mortal must have an end, poor Susie became sick unto death. The last I heard of the poor waif was, that he was seen with the dying object of his affection in his arms, hugging her and pressing her to his heart. The man was weeping like a child, and dreading the swiftly coming moment when he and his beloved Susie were to be parted."

There really must have been a substratum of good in this poor dissipated man, or he never could have cherished or inspired so sincere a love.

I have been rather uncertain as to the heading under which the following anecdote ought to be placed. As the reader will see, it illustrates reasoning and conscience, as well as the power of understanding human language. The last-mentioned attribute, however, being very strongly manifested, I have placed the story in its present position. I give the story in the words of the gentleman who kindly sent it to me:

"My grandfather, Mr. H——, of Gilchristcleugh, in Lanarkshire, possessed a watch-dog of the name of 'Help,' who was usually kept chained up. For some time repeated losses had taken place among the sheep, some of which were found torn and mangled, but only partially, if at all, devoured. Every effort to trace the secret enemy proved in vain. At last, while Mr. H—— was walking one day on the banks of a little river which flowed at the foot of the pasture hill, his attention was attracted by seeing his dog, whom he supposed to be safely chained near the house, running down the hill.

"As the dog drew near, it was seen that his mouth and fangs were covered with blood. My grandfather concealed himself so that he might watch unobserved what would follow. The dog walked into the river, dipped his face in the water, and shook his head backward and forward, until he thought that all traces of his guilt were removed. He then came out at the side next the house, toward which he proceeded, his master following at a little distance. The dog went to his kennel, and, with the help of his paws, put on his collar, which was lying with the chain on the ground.

"My grandfather walked up to him and said, 'Help, my poor fellow, there is no help for you.' He then went away, and gave the necessary orders for the dog's execution. But when the servant came to lead him to his doom the collar was once more empty, and 'Help' was never more heard of in the county."

It is plain that the dog must have perfectly understood the meaning, if not the exact words, of his master's speech.

No reproaches had been used; but he felt himself detected, and understood that he would have to suffer for his crime if he did not abscond.

Another story of a very similar character was

sent to me, but I have mislaid the MS., and can not remember the name of the narrator.

A gentleman had an old dog, which was so weighed down with the many infirmities of age that his master thought that the kindest treatment was a quick instead of a lingering death. Accordingly, he asked a medical friend to bring some poison. This he did, and, laying it on the table, said, without mentioning the dog's name, "That is the stuff which will do his business." The dog was at the time in the room; but soon afterward his master noticed his absence, and inquired about him. No one had seen the dog, and no one did see him again. In some mysterious way he had conjectured the object of the visitor, and had withdrawn himself, probably to die in some hidden spot, as is the way of all animals when they feel that the thread of life is being loosened.

Here I may observe that nothing would induce me to poison a dog, or allow it to be poisoned, except by a competent person who would administer a dose of prussic acid. Strychnine and arsenic, which are the usual poisons employed for killing dogs, cause horrible agony before death. Hanging and drowning are each objectionable, as the life is extinguished by degrees instead of suddenly, as ought to be the case. A bullet or a charge of shot through the brain is by far the most humane mode of destroying life, as the great centre of feeling is instantaneously crushed, and there is no time for even the slightest sense of pain.

In the story narrated below it is evident that the dog perfectly understood the words of his mistress, for he did violence to his own feelings, and obeyed the wish which his mistress conveyed in her rebuke. A thoughtless child, if reproved for a similar action and behaving in the same manner, would be held to have acted in a way that became a being possessed of an immortal soul.

A lady, who is a thorough appreciator of animal character, writes to me as follows on this subject:

"Dogs perfectly understand human language when reference is made to them, even though the words are not directed to the dog personally. If my little dog ever heard me make a plan in which he was to be left at home, while I was to go somewhere without him, 'Nettle' invariably set to work to counter-dodge me, and often got his own way in consequence. It was impossible to resist his queer, elfish determination."

We knew a dog named "Bijou," a thorough-bred Spitzberger. The house in which he lived was one of a terrace with a veranda running throughout its whole length, only separated by a wooden railing at each house. This veranda was Bijou's favorite resort; here he carried his chicken bones, and here contemplated the proceedings of his neighbors.

"One day a half-starved dog spied one of his bones lying about, carried it to the mat at the door of the next house, and began eagerly to devour it. Bijou saw the theft from the window, sprang out with a threatening growl, carried off the bone, and replaced it on his own mat.

"His mistress, who had been watching the action of the dogs, said to him, 'Oh, you greedy dog! You can eat no more, and that poor dog is starving.' Bijou at once picked up the bone of contention, carried it to his starving fellow, laid it before him, and retired to his own house, from the window of which he contemplated, with a benign aspect, the disappearance of the bone."

The following letter, which was sent to me by the Honorable Grantley F. Berkeley, illustrates the individuality of character to be found in dogs, the love which they bear toward their master, and their knowledge of man's language:

"There was in my pet greyhound 'Brenda,' there was in my dear lurcher 'Smoker,' and there is now in my dear lurcher 'Bar,' and in my three setters 'Chance,' 'Quail,' and 'Quince,' a *refinement* of feeling and sagacity infinitely beyond that existing in multitudes of the human race, whether inhabiting the deserts or the realms of civilization.

"I can not better define it than by saying that, if I give these dogs a hastily angered word in my room, though they have never been beaten, they will, with an expression of the most dejected sorrow, go into a corner behind some chair, sofa, or table, and lie there. Perhaps I may have been guilty of a hasty rebuke to them for jogging my table or elbow while I was writing, and then continued to write on. Some time after, not having seen my companions lying on the rug before the fire, I have remembered the circumstance, and, in a tone of voice to which they are used, I have said, 'There, you are forgiven.' In an instant, the greyhound Brenda would fly into my lap and cover me with kisses, her heart tumultuously beating. After she grew old, her joy at my return home after a long absence has at times nearly killed her; and when I was away, the bed she loved best was one of my old shooting-jackets, but never when I was at home.

"Had I time, I could look up many an instance of soul, in some senses of the word; but I have not."

Here is another of a physician's reminiscences, showing that dogs can exchange ideas with human beings, and understand the language of man:

"Having been much taken with the intelligence, faithfulness, and beauty of a terrier belonging to a coachman whose family I attended, I purchased three of her pups for myself and two friends.

"Shortly after this the coachman and his family moved to a new house, where they had no friends. His wife was taken suddenly and seriously ill, and could not stir from bed. The dog lay constantly at her feet, and never moved till the door was opened, when off she set in great haste. She went to the house of the suffering woman's parents, made a great and unwonted noise, and often went to the door, even laying hold of the woman's gown.

"The animal never rested until she followed it, when it manifested every token of approbation, as it looked around from time to time. It went straight to her daughter's house, when the real cause of the strange conduct of the dog was understood. The dog resumed its place, and scarcely left it for a moment until its mistress recovered. I then said, 'Now, "Missy," you may come with me,' when she went all her round with me, and returned home after dinner. This was repeated several times.

"The dog seemed really to understand many things you said, and even to forestall your wishes."

In this interesting story we find in an animal a singular aggregation of faculties which are held in man to belong to the immortal, and not to the mortal part of his being. There is reason, *i. e.*, the deduction of a conclusion from premises. There is the power of forming ideas and communicating them to man, and the capability of understanding man's language, and, as the writer says, even of anticipating the wishes of her human friends. Lastly, there is the intense love for her mistress, combined with the power of self-sacrifice, which enabled her to keep her irksome watch by the sick-bed while her instinct was urging her to take her accustomed exercise in the open air.

The cat which is mentioned in the following anecdote is the mother of Tiny, whose exploit with a lobster has been previously narrated. The writer is Lady E——. The reader will see that it illustrates two subjects: first, the fact that the cat understood human language; and next, that she could make her message understood by a human being.

"After my great loss, whenever I was left alone in the room, 'Rosy' usually placed herself on the table beside me, and watched my countenance most earnestly.

"About this time, my sister, who was living with me, had been some time in her room, and, wishing her to come to me, I said, 'Rosy, go upstairs and tell Augusta that I want her.' The cat immediately jumped from the table, ran upstairs to my sister's room, leaped upon the chest of drawers by which she was standing, and, putting her paw on Augusta's hand, mewed, then hurried down-stairs, mewing and looking around. This proceeding, Augusta could not mistake, was intended to call her down-stairs; so she followed, and asked if I wanted her, as Rosy had been to call her. Rosy appeared delighted at being understood, and purred with satisfaction."

CHAPTER VII.

MEMORY.

Memory *versus* Materialism.—Connection of the Brain with Memory.—The Workman and his Tools.—Memory in the Insects.—The Tame Butterflies.—Sir J. Lubbock's Tame Wasp.—Bees and Wasps find their Way by Memory, not by Instinct.—Comparison with Human Beings under Similar Circumstances.—Memory the Means by which Animals, as well as Men, are Capable of being Taught.—Two Tame Moles which would Come when Called.—A Partially Tamed Tiger-cat.—Memory in the Water-hen and Cockatoo.—The Power of Memory among Poultry.—Memory in the Ass.—"Donald," the Galloway, and his Long-lost Friend.—Memory of the Wolf.—Rarey *versus* "Cruiser."—Memory in the Cat.—"Fan," the Blenheim Spaniel.

I SHOULD think that Memory must be rather a hard nut for materialists to crack. What is that which survives, though every particle of the material brain has been repeatedly changed? What is that which more or less deeply receives impressions and retains them through a long series of years? And even when they seem to be forgotten, they are often but hidden behind a temporary veil, which at the touch of a passing scent in the nostrils, a dimly heard sound striking upon the ear, the waving of a branch, or the nodding of a flower, appealing to the eye, is in a moment rent asunder, and scenes long forgotten are reproduced before the memory as vividly as though time had been annihilated. Nothing is omitted; but there comes a minute and instantaneous insight into every detail, that gives us some faint idea of the omnipresence and omniscience of the Creator, to whom space and time are absolutely as nothing. For a moment we escape from our fleshly tabernacle, and we see and hear with our spiritual and not with our material organs of sight and hearing.

As to ourselves, we expect that we shall retain our memory, and carry it into the next world. We expect to recognize in the spiritual world those whom we have loved on this temporal world. Memory, therefore, must be spiritual and eternal; and wherever memory can be found, there is an immortal spirit. Apart from Revelation, which we have already considered, there is no stronger evidence of a future life of man than memory, and, in pure justice, if we apply this proof to ourselves, we ought to apply it wherever memory is found.

Some have said that memory is a mere emanation from the brain, and have tried to prove their point by asserting that which no one ever denied, that an inferior brain is coupled with an inferior intellect, that if the brain be injured by any cause all the powers of thought are weakened, and that if it be seriously damaged all powers of thought are utterly in abeyance.

All this is true enough, but it affords no proof that thought is the creation of the brain. On the contrary, the brain is the organ or instrument of the thought-power, and stands to it in the same relation that a tool does to a carpenter. However good an artisan a carpenter may be, he can not turn out good work with a blunt tool, nor any work at all with a broken one. So it is with the brain: it is but the tool of the spirit, and, if it be injured in any way, the keenest intellect will be unable to work with it.

Moreover, memory exists in creatures *which have no brain at all*. Take, for example, the insects, which have no real brain, but only a succession of nervous ganglia running along the body, and in many of them we shall find the faculty of memory very strongly developed.

Some ten years ago, I gave, in my "Glimpses into Petland," published by Messrs. Bell & Daldy, an account of two butterflies which had been tamed by a lady. One of the very critical weekly papers was good enough to treat the whole story with scorn and derision, saying that I gave it *as* from a lady, and thereby insinuating that the account was a willful imposition on the public. The story had been told to me by the lady in question, whom I have known for many years, and at my request she gave it in writing.

Here is the story, as published in "Petland:"

"Among the many pets that I have loved and lost, few have endeared themselves more to me than my butterflies, two of which I once kept for the space of a year and a half.

"They came into my possession when in their

chrysalis state, and I, not knowing any thing of entomology, shut them up for safety in a cabinet having glass doors. The cabinet stood near a small window in my bedroom. I was very unwell that winter, and therefore a fire was kept up in my room night and day. Therefore the room was very warm, and I suppose that the little butterflies were deceived thereby, and thought or dreamed that summer smiled upon the earth; for, a few days after Christmas, to my astonishment and delight, a little yellow butterfly was seen fluttering feebly within the cabinet.

"My attention was first diverted to the cabinet by the playful gambols of a pet pussy, who had mounted on a chair, and stood upon its hind legs, pawing at the little creature through the glass. I soon sent pussy away, opened the cabinet, and tried to induce the butterfly to alight on my hand. But it was either dazzled and bewildered at finding itself in its new and extended sphere of existence, or had already learned the fear of man; for at the approach of my hand it flew wildly about, and finally settled down as if exhausted.

"I now became most anxious to feed the little thing; but how this was to be achieved I had not the slightest idea, nor could any one in the house advise or help me in this important matter. Moreover, I was loudly ridiculed for the bare idea of trying to tame and feed butterflies.

"However, I remember that the poets all agreed in saying that butterflies sipped nectar from the opening flowers, and therefore turned my attention to the manufacture of a substitute for nectar; so, having obtained some honey, which I diluted with rose-water, I put one drop into the centre of the open blossoms of a fairy rose, and placed the little plant in the cabinet. I soon had the joy of seeing the little thing flutter around the rose, and finally settle upon it.

"Whether it really drank or not, I can not say. I thought that it must have done so, as it appeared to grow stronger and more lively every day. I fed it in this manner for a fortnight; and by the end of that time it became so tame that it would step off the flower or any thing else on which it might be standing, and appear quite happy and at rest upon my hand.

"It also appeared to understand that I wished it to come to me when I called it by the name of 'Psyche,' that being the name which I had given to the insect.

"About three weeks after the advent of Psyche we were gladdened by the addition of another butterfly to our establishment—a peacock. He was strong and vigorous from the first, and flitted swiftly about, like a beam of prismatic light. I used to fancy that they talked to each other, as he at once fell into the ways and habits of the other; and when I called Psyche, he too would come. I gave him another name, but he never seemed to understand that it belonged to him.

"They lived in this way until the earth had donned her glowing summer robe of lilies and roses, when I was told that their life-power could only extend over a month or two, and that it was cruel even to keep them as happy prisoners. I was therefore induced to give them their liberty. The cabinet was placed with open doors before the window.

"It was many days before the butterflies ventured to leave the window-sill, and this much to my joy, for I thought that it might be affection for me that held them back. However, one day, with many bitter tears, I saw them depart and join some wild companions; but at night we found them again in the cabinet.

"On the following morning they left us, and came not back again until the cold and stormy September weather set in.

"Yet, when in the garden, they would come if I called them, and rest for a short time on my hair or hands. At length, on a cold, windy day in September, we saw them on the window-sill, and, on our opening the window, they came in and resumed possession of their old quarters, and abode there for the winter.

"It is true they were but poor-looking objects to what they were when they went forth. The world seemed to have used them rather roughly, for the sheen had gone from the rich wings of the peacock butterfly, and the soft yellow bloom from Psyche's plumage. Nevertheless they were welcome guests; and, though ragged and wayworn, were not the less loved.

"We observed that during this winter they slept more than they did formerly. They also manifested pleasure when sung or talked to, and were very fond of being waved about and danced up and down in the air, while they would sit upon the hand quite calmly. I think that the movement must have reminded them of the nodding flowers and fresh breezes of their summer life.

"The sun and earth ran their appointed course until they brought us to another bright June, and again I bestowed the boon of freedom on our fairy pets, who went forth gayly; but, alas! never to return. One day, after a heavy thunder-storm, we found the inanimate form of a yellow butterfly upon the window-sill. I took it up lovingly, and did my best to revive it; for I believed it to be the material form of my own beautiful Psyche, who had sought refuge from the storm, but found the window closed. Of this I can not be sure, for all our efforts to restore her were in vain. The wondrous essence that had given it life, beauty, motion, affection, and memory had returned to the hand of its mighty Creator; and with him let it rest.

"The peacock butterfly never returned: whether he fell a victim to that aerial shark, the dragon-fly, or died of age, sickness, or forgot his early friends, I know not.

"I have since tried to tame other butterflies, but never was so successful, although I have taught three or four to know me and to come at my call. Indeed, circumstances have never been so favorable; for I never had any other butterflies in their chrysalis state, nor have a room and a cabinet been ready to receive them."

There are one or two points to be noticed in connection with this story. The first is, that the narrator, as she says, knew nothing of entomology. She was not aware that the yellow butterfly was our common "brimstone," and the so-called peacock butterfly was in reality a "small tortoise-shell," these being the earliest and the hardiest of British butterflies, the "brim-stone" being almost invariably the first butterfly to be seen, while the "small tortoise-shell" follows it after a short interval. I was much puzzled at the description as given in writing, and it was only by getting a description of the so-called "peacock" butterfly *vivâ voce* that I was able to identify the insect. She did not know how a butterfly fed itself. She knew nothing of the hibernation of these insects, and yet if a practiced entomologist had written the story, it could not have been more accurate in these scientific details.

But if the reviewer will not believe the account written by a lady, although authenticated by myself, he may believe Sir John Lubbock's account of a far more difficult task, namely, the successful taming of a wasp. Here is the story in his own words:

"DEAR SIR,—In answer to your inquiries, I beg to send you the following particulars about my poor wasp:

"I took it, with its nest, in the Pyrenees last May. The nest, which was beautifully regular, consisted of about twenty cells, the majority of which contained an egg; but as yet no grub had been hatched out, and, of course, my wasp was as yet alone in the world.

"I had no difficulty in inducing her to feed on my hand; but at first she was shy and nervous: she kept her sting in constant readiness, and once or twice in the train, when the official came for tickets, and I was compelled to hurry her back into her bottle, she stung me slightly—I think, however, entirely from fright.

"Gradually she became quite used to me, and, when I took her on my hand, evidently expected to be fed. She allowed me to stroke her without any appearance of fear, and for some months I never saw her sting.

"When the cold weather came on, she fell into a drowsy state, and I began to hope she would hibernate and survive the winter. I kept her in a dark place, but watched her carefully, and fed her if ever she seemed at all restless.

"She came out occasionally, and seemed as well as usual till near the end of February, when one day I observed that she had nearly lost the use of her antennæ, though the rest of her body was as usual. She would take no food. Next day I tried again to feed her; but the head seemed dead, though she could still move her legs, wings, and abdomen. The following day I offered her food for the last time, but both head and thorax were dead or paralyzed; she could but wag her tail—a last token, as I could almost fancy, of gratitude and affection. As far as I could judge, her death was quite painless; and she now occupies a place in the British Museum."

The reader will see that, in both these examples of tamed insects memory was absolutely indispensable, and that without the existence of this faculty it would have been impossible to influence them with human ideas.

As to the wasps, the late Mr. Stone, who had made them his special study, told me a portion of these insects' life-history which proves the existence of memory. We were speaking of the "homing" faculty of various animals, especially pigeons, bees, and wasps, and were debating whether the faculty were due to instinct or reason. Mr. Stone gave his decided opinion that all those creatures were guided by reason, the insects as well as the birds.

He said that any one who was accustomed to the ways of these insects could tell by the manner in which a wasp left the nest whether it were an old or a young one. An old wasp crawls to the entrance of the nest, and at once darts off without any ceremony. A young one, however, when going out on its first expedition, acts in a very different manner. When it has emerged from the entrance, it turns round and examines the spot; it then takes to wing, but flies backward and forward in front of the nest, and always looking toward it as if taking notes of the bearings, and gradually increasing its distance until it is out of sight.

Here, then, is a distinct exercise of memory as well as of reason, the creature impressing on its mind the appearance of the objects near its nest, and acting on the result of those impressions. Human beings act in just the same way when traversing for the first time a locality through which they will have to return. And yet, as I have already stated, the wasp has no true brains.

Mr. Stone remarked that he has seen bees act in a similar manner when their hive has been moved to a spot at any distance from that which it formerly occupied.

We will now pass to some of the higher animals, taking, first, one or two examples of creatures that are not usually subject to domestication, and can therefore have received no teaching by means of their parentage.

By means of this faculty almost any living being is able to be taught by man, while, if memory were absent, no teaching would be of the very slightest use. The mole, for example, seems to be about as difficult a subject as can well be imagined, and yet I knew of one case where a mole was perfectly domesticated, and another in which it was partially tame.

The former was a specimen of the albino, or white mole, a variety which is tolerably common. It was living at St. Malo, in 1856, and the story of its life was told to me in 1857.

It knew its name, would come to its master when called, and had learned to perform some little tricks: for example, when told to do so, it would tumble over on the table, and would fetch coins if they were scattered within its reach. The animal had a curious preference for silver over copper coins, probably because the sensitive nerves of the mole were affected by the copper. It always ran faster, and seemed more pleased when it had a silver coin in its mouth than when it had been dispatched after a copper coin. What it might have done in these days of the light bronze coinage, I do not know.

The second example of a tame mole was one of the common brown animals, which had got into a garden, and was doing much damage. The gardener, being practical and not æsthetical in his tastes, did his best to kill the mole; but the inhabitants of the house, being æsthetical rather than practical, tried to tame the animal, in which they partly succeeded.

The aperture by which the mole usually came into the open air was situated under a sage-bush, and near the opening a piece of raw meat was laid. The delicate organs of the mole soon perceived the supply of food, and the animal, after he had finished his meal, came to look for some more. This was given him, accompanied by the sound of his name, "Barty," an abbreviation of Bartimeus. Sometimes the mole was too far away from the aperture to hear his name, and in these cases a measured stamp upon the ground was sure to bring him to his meal.

Here is an instance of the influence of memory upon an animal which is not often tamed, and which in this case happened to be a peculiarly fierce and sullen individual.

Some years ago I was a constant visitor to the Zoological Gardens, and used to make acquaintance with the various animals as far as they would allow me to do so.

One day I was struck with the beauty of a very large and beautiful ocelot, or tiger-cat; but the animal seemed to be a new-comer, and was very wary and fierce, declining to respond to any overtures that were made. At last, when standing by the cage on a hot summer's day, I thought that I saw a mode of getting at the animal's feelings. The place quite swarmed with flies, mostly blue-bottles, a few of which occasionally got inside the bars of the cages. Seeing the ocelot try to catch one of the flies, I captured a fine large blue-bottle, and held it close to the bars so as to make it buzz, and waited quietly. After a while the ocelot came cautiously up, and, after one or two feints, took the insect and ate it. I immediately caught another, and offered it in the same way, giving a low whistle at the time. This time the ocelot took it without much difficulty, and in half an hour or so he came at once to the whistle, and took the fly.

On the next visit I repeated the proceedings, the ocelot perfectly recognizing me; and after one or two visits, the beautiful creature would press itself against the bars to be caressed, and to have its nose and chin rubbed, just as does a favorite cat. The keeper happened to come in while I was talking to the ocelot, and was quite alarmed, saying that even he did not dare to trust his fingers between the bars. Now the keepers are specially kind and gentle toward the animals under their charge, and can do wonders with the fiercest of animals; so that for a keeper to be unable to trust his hand in a cage, shows the ferocity of the animal confined in it. I fully believe that in this, as in most other cases where an animal is ferocious, fear, and not ill-temper, is the real cause of its conduct.

The following account of a tame water-hen was sent to me by the owner of the house—a lady well known in the literary world:

"Some five or six winters ago two water-hens made their appearance in the mountain brook which runs through our lawn, and were constantly to be seen upon the grass. One was larger than the other, of a deeper color, and we supposed them to be a pair. The winter was exceptionally severe; there was more snow than usual, and when it melted the smaller of the two was found dead. The other remained until March, when it disappeared. During its stay it had learned to come toward the dining-room window while the pea-fowl were being fed, and, if food were thrown to a little distance, would pick it up.

"The second week in the October following it again made its appearance, and remained through the winter, becoming tamer every day. At last, whenever it heard the window opened, it would hasten, half running, half flying, to be fed. Every year it has appeared and disappeared with as much regularity as the swallows, and always about the same day of the same month. Now as soon as it arrives it is perfectly tame, and comes running up as soon as the sound of the opening window is heard.

"I always feel sorry when the time of its departure arrives, and gladly welcome its return. It has never had a companion, but it must leave for the purpose of getting a mate. Yet it never brings one here, nor have I ever seen another water-hen within miles of this place."

Here is a good example of memory in the case of a domesticated bird:

"Our noble yellow-crested cockatoo was the especial pet of the eldest daughter of the house. The young lady married an officer, and was absent from the old house for nearly three years. Her anticipated advent on a visit to her father was of course talked about, and we may imagine the cockatoo pricked up her ears at the sound of her name. The moment the carriage stopped at the door she flew down from her perch, and, before mamma or sister could greet her arrival, was outside the front door with 'Kiss me, my dear; kiss me, Sa; kiss me, Sa.'"

Why it should be I can not tell, but our domestic poultry are sadly neglected in the way of human education; and yet that they are perfectly capable of receiving it, if properly given, I am quite sure, having seen many instances in which poultry of various kinds have preferred the companionship of man to that of their own kind. I knew personally a chicken and a duck who entirely repudiated their proper companions and domicile, preferring men to birds, and the drawing-room to the poultry-yard. The chicken had been an ailing little creature; and being carefully tended until its restoration to health, attached itself vehemently to its nurse, and used to follow her over the house, calling her anxiously until seated in her lap. I shall presently have to tell several anecdotes of poultry, but under a different heading, so confine myself to one which was sent to me by the chief actor:

"I am no poultry-fancier, being perfectly ignorant of the distinction between Brahmas, Cochins, etc. We have only a few fancy bantams.

"During the last illness of a favorite riding-horse I was a frequent visitor to her stable; and one wintry morning, after a snow-storm, one of these tiny bantams looked so cold and pitiful that I put it on my hat, and thus transferred it to the warm stable. I never could find much intelligence in the poultry tribe; but this little bird, which I named 'Jemmy,' found the climate of the stable so enjoyable that, in order to obtain an entrance, it watched my visits, always flying up to my hat directly I approached.

"Mimicry is the gift of monkeys, but I know that fowls are endowed with it. Jemmy had some little brothers and sisters, who followed his example. Not wishing to accommodate the whole family on my hat, I made it my custom to push the others off. Once, by mistake, I pushed off Jemmy, who made me aware of the fact by a great cacophony, and resented my unintentional rudeness to such an extent that it was many days before I could obtain his forgiveness, and induce him to resume his high position. Once I entered while wearing a bonnet: his efforts to obtain his usual comfortable footing were most absurd, and at last he descended in great disgust at the alteration."

As for anecdotes of the domesticated animals, such as the dog, the cat, the horse, and the ass, there are so many that I am obliged to restrict myself to a very few. Indeed, every one who has had personal experience of these animals must have remarked the great strength and endurance of their powers of memory.

The following story is by the late Rev. Cæsar Otway, and is told in his lecture on the "Intellectuality of the Domestic Animals:"

"I shall tell you what I know of an ass. There is a lady resident in a parish where I was for some years minister. She is the most tender-hearted of the human race; her tenderness, though a general feeling, is principally confined to the lower animals. I am disposed to think that if in India or Turkey, she would leave all her worldly goods to endow a hospital for deserted, disowned, and abused animals.

"Well, this lady was walking along the road,

and she met a train of tinkers proceeding toward Connaught, and one tall, tan-skinned, black-haired, curly-polled fellow, in all the excited cruelty of drunkenness, was belaboring his ass's sides with a blackthorn cudgel. This was too much for my friend. She first rated the man for his barbarity: she might as well have scolded Beelzebub. She then coaxed the ruffian, and asked him if he would sell the creature, which he consented at once to do, asking, of course, three times the proper price. You may judge of the joy of this amiable woman when the beast, now her own, was relieved from its panniers, allowed to roll about in the dust, and graze at liberty.

"For a long time she kept him perfectly idle, until he recovered his spirits; then he became troublesome, and would break his bonds, and used to go a-braying and curveting, and seeking for asinine society all over the country. Idleness is certainly, after all, a bad thing for asses as well as men, and so this capricious fellow found it; for shortly a tinker, perhaps the very one that sold it, stole it, and for three or four years there were no tidings of the ass, until one day, as his kind mistress was taking her usual walk along the road, she saw a man urging along an ass straining and bending under a very heavy cart.

"Now the moment my friend came near there was an alteration in the deportment of the ass; immediately the ears that were but just now hanging listlessly over its eyes were cocked, and its head elevated in the air; and, raising its voice more like a laugh than a bray, it urged itself under its heavy load into a trot, and came and laid its snout on the shoulders of the lady, who at once, and not until now, recognized her long-lost property, which she had again to purchase at a high price. It is many years since that occurred; the beast is alive, and so is the lady. I hope it won't be her lot to see in it that rare spectacle, a dead ass."

An adventure of a nearly similar nature occurred to the gentleman who furnished the account of the miller's dog at Maxwellheugh, and who has kindly taken a great interest in the object of this work:

"When I was a boy, my father bought from a neighboring farmer a gray Galloway pony, who was very vicious to all with whom he came in contact except myself. The way in which I acquired so much power over him was by feeding him with bread, and showing him other acts of kindness.

"Some years afterward I left home, and when I returned to my father's house I found that 'Donald' had been sold, and that all trace of him had been lost for about seventeen years. At that period, being resident in a village in a neighboring county, I saw an old white horse in a cart, and thinking that it might be the same animal, I went up to him in the same way as I used to do in boyhood, and said, 'Donald.' He immediately turned his head to me, laid it on my shoulder, pawed the ground, rubbed his nose upon my arm, and showed the greatest possible affection.

"The driver of the cart came out of a shop, and warned me to keep away from the horse or he would bite me. I moved up the street, when Donald became restive, wrenched the reins out of the lad's hands, followed me along the street, and it was not until I entered a house that, after much difficulty, he was induced to move away."

This is a really wonderful act of memory on the part of the horse, and not at all a bad one on the part of the man; and the incident affords a direct proof that memory is a common possession of man and beast. That the man should recognize the animal which he loved in his boyhood was a tolerably fair exercise of memory; but that the horse should recognize the man is even more astonishing. From boyhood to manhood, the lapse of seventeen years makes such changes in personal appearance that, as a rule, the man of thirty can scarcely be recognized even by those who knew him well as a boy of thirteen. Nor can the voice give any help in recognition, for the deep tones of the manly voice are as unlike the shrill sounds of a boy's "treble pipe" as is the bearded face of the man to the smooth cheek of the boy.

Dress also makes a great difference in the appearance of a human being; and when we consider that the dress of a man is quite unlike that of a boy, we must appreciate the strength of memory which enables the horse to recognize his friend in spite of so many alterations.

Anecdotes of a similar character are plentiful, and even the wild beasts are known to remember a human friend after a long lapse of years. In Hardwicke's *Science Gossip* of October, 1871, there is an account by Mr. W. W. Spicer of a wolf at the Zoological Gardens at Clifton with which he struck up a friendship. He was forced to leave Clifton for some two years, and on his return went to see his friend:

"I at once set to work to test the wolf's affection and retentiveness of memory by whistling in a low tone at as great a distance from the den as allowed my watching its movements. At the first sound, the animal, which before was 'loafing' about in a listless manner, raised its head and listened; and on my continuing to whistle, it bounded against the bars with every mark of joy.

"Long before I reached the cage he recognized my footsteps, and strove to engage my attention by whining and throwing himself into all kinds of queer positions. My welcome, in fact, was of the warmest kind, and I left him with, I was going to say, mutual expressions of sincere regret; for if ever an animal gave expression to its feelings, it was this poor wolf, who recognized me after so long an absence."

These anecdotes fully corroborate the opinion which I have always held with regard to the relationship between man and beast. The latter was intended to serve the former, and there is nothing in the hands of man half so powerful in educating the lower animals as thoughtful kindness. Inflexible decision, combined with gentleness and sympathy, are irresistible weapons in the hand of man; and I do not believe that there is any animal which can not be subdued if the right man undertakes the task. By this mixture of firmness and kindness that raging wild beast of a horse, "Cruiser," was in three hours rendered gentle and subservient, obeying the least sign of his conqueror, and allowing himself to be freely handled without displaying the least resentment.

I once saw Mr. Rarey operate on a splendid little black Arab horse that flew like a tiger at him, kicking, biting, and screaming at once, now attacking with his jaws, and now with his heels. He might as well have attacked his own shadow; for, just as the Spanish bull-fighter absolutely plays with the furious beast in the circus, so Rarey seemed to play with the animal, stepping quickly on one side as it made its rush with open mouth, and then, as it spun round and lashed out with its heels, being on one side, just out of reach. Within half an hour Rarey and the horse were lying together on the ground, Rarey's head resting on one of the hind hoofs, and the other hoof being laid on his temple. He then got up, mounted the animal, dismounted by sliding over its tail, and finally, with hands in his pockets, ran round the circus, the horse's nose resting on his shoulder. He had impressed upon the animal's memory that no harm was intended; and so the horse, instead of feeling fear and anger, conceived an affection for the man who inflicted no pain, and yet showed that he must obeyed.

The following anecdote of a cat demonstrates several traits of character which are common both to man and beast. I was rather doubtful under which head it should be classed; but as it illustrates the present subject, I have placed it here:

"I confess myself a great friend and admirer of horses and dogs, but care little for cats in general, although, when away from home pets, I often make playthings of them. Did you ever know a landlady without a cat, visible or invisible? We had rooms in Berkshire, and the morning after our arrival, on entering the dining-room, I saw a real, visible cat sitting on our breakfast-table, and reducing the quantity, if not the quality, of the milk. The milk-pot being narrow at the top, she obtained it by putting in her paw, curling it round, and then lapping it up. (Animals are never afraid of me, nor do I wish them to be so.) I allowed Puss to continue her depredation on the milk; we breakfasted without it, and her theft remained unpunished.

"After we had been there some time, Puss listened every morning until I rang for the tea-kettle, which she always accompanied to the breakfast-table. One morning I was later than usual, and while dressing I was surprised to hear the cat mewing at the bedroom door. As she had not before done so, I let her in, with the remark, 'Puss knows I am late, and is waiting for her breakfast.' I was, however, quite mistaken: she was too ill to eat, but came to me for that sympathy which she could not obtain from others. After our departure no one cared for the poor animal; she was first neglected, and then killed for being delicate. She was such a gentle and affectionate creature that I would have taken her to my own home if I had known her impending fate."

Here are several mental characteristics exhibited by the same animal. Her reason taught her to get the milk out of the jug with her paw when she could not reach it with her tongue. I know a very intellectual cat, the grandmother of my own lamented "Pret," who would steal bottled porter in the same way. She would not take milk, but porter had a fascination that she could not withstand. Then this cat's memory retained the recollection of kindly treatment; and so she not only became partaker of the daily meal, but asked and obtained loving sympathy when she felt herself ill. A child who had been kindly treated would have acted in precisely the same manner.

Another instance of reasoning and memory brought to bear on sickness has been communicated to me by a friend:

"As illustrative of memory, take the following anecdote: A pet Blenheim spaniel, 'Fan,' had had two or three litters. At the birth of the third or fourth family (who were all dead born) she was extremely ill for some days, refusing every thing in the shape of nourishment, till by dint of much coaxing and petting she took captain's biscuits, and lived on them solely till quite convalescent. In succeeding accouchements she refused all other food till her master thought of the biscuits. When offered, they were immediately eaten with avidity, and she kept to that particular diet forever afterward when nursing."

In all these examples of memory, the reader will probably have remarked that there must be something more in this faculty than a mere production of a material brain. In several cases there was no brain at all; and in others, where a brain did exist, its material particles must have been repeatedly changed, while the ideas impressed upon the memory still remained in full force.

CHAPTER VIII.

GENEROSITY.

Different Senses of the Word Generosity.—Firstly: the Sense of Liberality.—Two Grateful Cats.—"Pret" and his Mice.—Pret Entertaining his Friends.—Generosity Before Justice.—Another Grateful Cat.— The Cat Providing Mice for Two Sea-gulls.—The Retriever "Nellie" and her Cat Friend.—"Barbekark," the Esquimau Dog, and the Reindeer.—His Self-denial and Power of Command.—The Generous Sparrow.—Secondly: the Sense of Magnanimity, or Indisposition to Resent an Injury, though Possessing the Power to do so.—Might and Mercy Convertible Terms.—Anecdote of Cribb, the Prize-fighter.—"Lupo" and his Little Friends.—A Dog Fighting Himself into Favor.—My Bull-Dog "Apollo" and the Retriever.

IN whatever sense we take the word which heads this chapter, *i. e.*, whether we accept it as signifying liberality or magnanimity, the quality is acknowledged to be a very lofty one, and one which infinitely ennobles the characters of those who possess it.

Take the former sense of the word.

It is, in fact, an attribute of God himself, who gives us freely all that we possess, and so sets us an example of generosity to our fellow-creatures. That we recognize this as a fact is shown by the extraordinary number of disparaging epithets and nicknames which are employed in designating those human beings in whom generosity is more or less wanting. Miser, skinflint, churl, screw, muck-worm, curmudgeon, scrimp, lick-penny, etc., are among the nicknames bestowed upon such unfortunate persons; while among the epithets are such flowers of speech as stingy, shabby, mean, parsimonious, hard-fisted, sordid, covetous, niggardly, and a host of similar terms too numerous to mention.

Now if it be admitted that the possession of generosity ennobles man's character, while the lack of that quality debases it, we ought not to deny the plain inference that when we find a beast possessing generosity, and a man devoid of it, the beast is in that particular not only the equal but the superior of the man. And that generosity, being a divine attribute, belongs to the spirit and not to the body, I should presume that no one who believes in Christianity is likely to deny; so that wherever we find this characteristic developed we must admit the presence of an immortal spirit.

I will now produce a few authenticated anecdotes in order to prove that the lower animals do possess generosity in the sense of liberality, several of the circumstances having occurred within my own observation, and the others being authenticated with the names of the writers.

With regard to the sense of generosity and gratitude which can be developed in the cat, the following anecdote was related to me by a friend of the owner of the animal:

The cat had some kittens, and one of them was taken ill, and was apparently in a dying state. The mother did all that she could for it; and then, finding all her efforts useless, brought the sick kitten to her mistress, laid it in her lap, and left it in her care. The lady accepted the charge, nursed the kitten through its illness, and at last was able to give it back to its mother quite restored to health.

Some time afterward the lady herself was seized with illness, and was unable to leave her bed. By some mysterious means, whether by mere instinct or by gathering the meaning of the conversations around her, the cat became aware of her mistress's illness. Finding herself unable to enter the room by the door, she contrived to climb up the wall of the house, scrambled in at the window, jumped on the bed, and laid on the pillow a mouse, as an offering of affection and gratitude.

Since I began to write this book I have received many anecdotes of a similar character, and in nearly all, if not all of them, *gratitude* was the existing cause of the animal's generosity. Indeed, I could easily have made a separate chapter on the subject; but not wishing to multiply chapters, I have included them under the present heading. Here is a story which appeared in *Good Words* for December, 1873:

"A cat in a Swiss cottage had taken poison, and came, in a pitiful state of pain, to seek its mistress's help. The fever and heat were so great that it dipped its own paws into a pan of water—an almost unheard-of proceeding in a water-hating cat. She wrapped it in wet linen, fed it with gruel, nursed it, and doctored it all the day and night after. It revived, and could not find ways enough to show its gratitude. One evening she had gone up-stairs to bed, when a mew at the window roused her. She got up and opened it, and found the cat, which had climbed a pear-tree nailed against the house, with a mouse in its mouth: this it laid as an offering at its mistress's feet, and went away.

"For above a year it continued to bring these tributes to her. Even when it had kittens, they were not allowed to touch this reserved share; and if they attempted to eat it, the mother gave them a little tap—'That is not for thee.' After a while, however, the mistress accepted the gift, thanked the giver with a pleased look, and restored the mouse, when the cat permitted her children to take the prey which had served its purpose in her eyes.

"Here was a refined feeling of gratitude, remembered for months afterward, quite disinterested, and placed above the natural instincts (always strong in a cat) toward her own offspring."

Urged by a similar feeling, my own cat, "Pret," used invariably to give his mice to me.

He used to kill the animal in a most curious manner, *i. e.*, by taking it, while quite unhurt, by the tip of the tail, carrying it to the top of the house, and dropping it down the well of the staircase. After repeating the process a few times, he would bring the mouse to me, and while I stroked and praised it, would keep rubbing himself against me and purring his content. He then took the mouse again, played with it for a while, and then brought it back to me. If the study door were closed, and he could not gain admittance, he always left the mouse on the mat, previously having bitten off the animal's head. He had a strange fancy, also, for putting the mice into my bed; and once, on leaving my room in the early morning, I found no less than nine mice laid in a row just outside the door. Afterward when we moved into the country, and he took to catching rats instead of mice, he acted in precisely the same manner, sometimes bringing me three or four rats in a single day.

Now in both these cases the motive was one that would show credit to humanity. There is nothing that cats like so well as a mouse, and yet, just because they thought mice the most precious object in the world, the cats gave their mice to those whom they loved. Affection, self-denial, generosity, and gratitude were thus exemplified, all being qualities which of necessity belong to the spiritual and not to the animal nature.

Pret was also remarkable for generosity toward his own kind. An example of this trait of character is given in my "Glimpses into Petland," published by Messrs. Bell & Daldy. The animal was then living in London:

"When he was a few months old he began to scrape acquaintance with other cats, and used to meet them in a back-yard, which, by common consent both of cats and householders, seemed to be the feline club-house of the neighborhood. Now it was very well of Pret to be social in his habits, but when he took to hospitality the question became serious.

"It is true that he never allowed strange cats, no matter how big they might be, to enter the house; but then he was fond of entertaining his friends in the yard, and was in the habit of bringing his dinner to the club for the benefit of his acquaintances, and then wanting a second dinner on his own account in the evening. He even went so far as to be disgusted with the meals furnished to a neighboring cat, thinking that cat's meat was not fit for feline consumption. Acting on this supposition, he was seen to take away the cat's meat as soon as it was brought by the itinerant purveyor, to carry it into the cellar, bury it under a heap of small coal, and to take his own dinner up-stairs for his friend.

"Even these proceedings might have been pardoned; but Pret's generosity developed so rapidly that we should have been obliged to devise some effectual check, had not a removal to another house put an end to the acquaintance.

"Finding that his own meals were not sufficient to entertain his friends in the liberal manner in which he thought himself bound to act, he took to ransacking the larder, into which chamber he contrived to gain admission in spite of many precautions. In vain did we keep the doors shut and the windows fastened, so as to exclude any animal larger than a mouse, for Pret always managed to enter the forbidden precincts whenever he chose. At last we found out that he achieved the feat by hiding under the servant's dress, and stealthily creeping in when she had occasion to visit the larder.

"On one occasion I heard an odd sort of a bumping sound on the stairs, as of some one who was dragging up a burden which could with difficulty be lifted. On going to investigate the source of the unwonted sounds, I found that they were caused by Pret, who had made a raid on the larder. He had contrived to drag out of the dish, and half-way up-stairs, the entire bone of a leg of mutton, resting on each step in order to get his breath, and then hauling the bone up the succeeding step. The ant pulling a stick over rough ground presents an exact resemblance to Pret dragging the heavy bone up-stairs."

It must be remembered that this labor was not undertaken for his own selfish purposes. He had not the least idea of eating the meat which he was carrying off, but intended to give it all to his friends.

An anecdote, curiously similar to that which has been related of two cats, has just been sent to me.

There was a cat whose kittens had been rescued from danger by her master, to whom she formed a devoted attachment. During his last illness she never left his room except for the purpose of procuring food, and even then she ate it hastily, and rushed up-stairs again as fast as possible. One day, in order to show her gratitude and affection, she went and caught a mouse, which she laid on his pillow.

I have always thought that the good qualities of the cat have seldom been appreciated at their just worth. This one trait of generosity, which we all agree in considering as one of the noblest characteristics of man, is developed very greatly in the cat, which, instead of being a greedy, selfish animal, as we are generally told, is really a very unselfish and generous one, capable of great self-sacrifice, and for objects which appear hardly worthy of it.

The following anecdote of generosity in a cat was told to me by Mr. Zwecker, the well-known artist, to whom I am indebted for so many admirable illustrations.

A friend of his had a couple of tame sea-gulls which ranged the garden freely, one wing of each being clipped, to prevent them from flying away. He had also a fine young cat, which struck up an oddly assorted friendship with the gulls. After a while she evidently compassionated their crippled condition, and thought that it prevented them from hunting. So she set to work at hunting for them, and was in the habit of bringing them little birds and mice, which they ate with the solemn satisfaction of a gull at meals. It is astonishing, by the way, what a large morsel a gull can swallow. I have often seen a gull take a large slice of bread-and-butter by the middle, and, in spite of the narrowness of its beak, the bird contrived to swallow the slice without putting it down or breaking it.

The following account of generosity on the part of a cat was sent to me by a lady living near Brighton. I knew both the animals mentioned. "Nelly" was a large, black, silken-haired retriever, and a great favorite.

"In the hot weather our large dog Nelly, whom you admired so much, used to be chained under a large oak in the grounds at the back of the house, just within sight of her kennel and the yard door. This was done that she might have the comfort of the cool position during the heat of the day, and at the same time command the back entrance to the house. This, however, took her away from the neighborhood of the cook, and the little scraps and dainty bits which used to be given to her now and then while the different meals were in course of preparation.

"At the same time, we had a dear, motherly old cat, who did not approve of the change of position in which her friend Nelly was placed. Still less did she approve of the cook putting all the scraps in a plate, instead of giving them to Nelly. So she set herself to work at conveying them to her friend, and every thing that was not too large for her to carry or drag along she took to the dog under the tree, and seemed delighted when she saw her friend eat them.

"Now she never stole any thing for herself, but she would always do so for any of the dogs. She used to carry little treats to a small dog that was chained up in the house, but this was after she developed the plan of helping Nelly to the dainties of which she, in her pussy-cat brain, considered her friend to have been defrauded."

I know of a somewhat similar case, where a cat was seen to steal a piece of meat and run off with it. She was followed, and then it was found that she had stolen the meat in order to feed a miserable cat that had fallen into a deep hole, and could not get out.

The late Captain Hall, author of the well-known "Life among the Esquimaux," was a great appreciator of the lower animals, especially of the dog. There was one sledge-dog in particular who was a particular favorite with Captain Hall, in consequence of his intellectual character, and the odd, quaint ways which he had. The reader will admire the singular self-denial and generosity of the animal as shown in the story told by Captain Hall:

"As Koojesse cautiously proceeded, we all watched him most eagerly. Fifteen minutes saw him 'breasted' by a small island, toward which the deer approached. When they were within rifle-shot he fired, but evidently missed, for the game wheeled around and darted away.

"Directly the report was heard, 'Barbekark,' my Greenland dog, bounded off toward the battle-ground, followed by all the other dogs. This was annoying, as it threatened to put an end to any more firing at the game; and, if they would have heeded us, we should instantly have recalled them. But it was now useless. The dogs were in full chase, and fears were entertained that if they got too far away, some, if not all of them, would be lost. At length we saw Barbekark, pursuing, not in the deer *tracks*, circuitous, flexuous, mazy in course, but in a *direct* line, thus evincing a sagacity most remarkable. The other dogs, not taking the same course, soon fell behind.

"On and on went Barbekark straight for a spot which brought him close upon the deer. The latter immediately changed their course, and so did Barbekark, hot in pursuit after them. Thus it continued for nearly two hours; first this way, then that; now in a circle, then zigzag; now direct, then at right angles, among the numerous islands at the head of the bay.

"For a while nothing more was thought of the affair, save an expression of regret that the dogs would not be able to find their way home, so far had they been led by the enticing game.

"A little before twelve, mid-day, Barbekark was seen coming back, and presently he came on board, with blood around his mouth and over his body. No importance was attached to this beyond supposing that he had come into collision with the deer; but as for killing one, the thought was not entertained for a moment. Those who had often wintered in the Arctic region said they had never known a dog to be of any use in hunting down deer, and therefore we concluded that our game was gone. But there was something in the conduct of Barbekark that induced a few of the men to think it possible he had been successful. He was fidgety, and restlessly bent upon drawing attention to the quarter where he had been chasing.

"He kept whining, and going first to one and then another, as if asking them to do something he wanted. The captain even noticed him jumping about, and playing unusual pranks, running toward the gangway steps, then back again. This he did several times, yet no one gave him more than a passing notice. He went to Keeney, and tried to enlist his attention, which at last he did so far as to make him come down to me.

"I was writing in my cabin at the time, and mention it; but I gave no heed, being so much occupied with my work. Perhaps, had Barbekark found me, I should have comprehended his actions. As it was, he failed to convey his meaning to any body. Presently one of the men, called 'Spikes,' went off to the wreck of the *Rescue*, and Barbekark immediately followed; but, seeing that Spikes went no farther, the dog bounded off to the northwest, and then Spikes concluded that it was really possible that Barbekark had killed the deer. Accordingly he returned on board, and a party of the ship's crew started to see about it, though the weather was very cold and inclement. They were away two hours; and when they came back, we could observe that each was carrying something like a heavy bundle on his head.

"Still we could not believe it possible that it was portions of the deer; and only when they came so near that the strange fact was perceptible, could we credit our senses. One man, almost Hercules-like, had the skin wrapped around him, another had half of the saddle, a third the other half, and the rest each some portion of the deer that we had all especially noticed. In a short time they were on board, and deposited their loads triumphantly on the scuttle-door leading to the cooking department below.

"Every officer and man of the ship, all the Innuits and Innuit dogs, then congregated round the tempting pile of delicious fresh meat, the trophy, as it really proved, of my fine Greenland dog, Barbekark. The universal astonishment was so great that hardly a man of us knew what to say. At length we heard the facts, as follows:

"Our men had followed Barbekark's return tracks for about a mile from the vessel, in a direct line northward; thence westward some two miles farther to an island, where, to their surprise, they found Barbekark and the other Greenland dogs seated upon their haunches around the deer lying dead before them.

"On examination, its throat was shown to be cut with Barbekark's teeth as effectually as if any white man or Innuit had done it with a knife. The windpipe and jugular vein had both been severed; more, a piece of each, with a part of the tongue, the skin and flesh covering the same, had actually been bitten out. The moment Sam, one of the men in advance of the rest, approached, Barbekark jumped from his watchful position close by the head of his victim, and ran to meet him, with manifestations of delight, wagging his tail and swinging his head about. At the same time he looked up into Sam's eyes, as if saying, 'I've done the best I could; I've killed the deer, eaten just one luscious mouthful, and lapped up some of the blood. I now give up what you see, merely asking for myself and these my companions, who have been faithfully guarding the prize, such portions as yourselves may disdain.'"

It is impossible that human beings could have acted more generously, and it is tolerably certain that few savages would have done as much. Indeed, after reading the accounts of the African savages as written by Livingstone, Baker, Grant, Burke, Burton, and other modern travelers, we can but come to the conclusion that if a number of savages in the service of a traveler had killed an animal, very few minutes would have elapsed before the carcass was torn to pieces.

See, also, how many human attributes are here shown. There is Reason. The animal, on hearing the gun, and seeing the deer go off, thought that his help was wanted, and at once gave it, with the assistance of his comrades, over whom he evidently exercised the authority that is so often evinced when peculiarly intellectual animals are brought in contact with those less highly gifted. The latter at once acquiesce in their own inferiority, and submit to the leadership of their acknowledged superior. His reasoning powers were again shown by the way in which he led the chase of the deer—not following their circular tracks, but cutting across them, just as if he had been a mathematician who knew that the chord was shorter than the arc.

Having killed the deer, he set his companions to watch the carcass, while he went off to bring assistance in carrying the deer home. He knew that although he and his companions could not get the deer to the ship, the men could do so, and accordingly he went to ask their aid in his own doggish language. He must also, before he started, have told his companions that they must not eat the deer. The generosity displayed by all the dogs is really wonderful, when we come to consider the circumstances. An Esquimau sledge-dog is always hungry; for, in the first place, the constant and severe work in which they are engaged is enough to give them a ravenous appetite; and in the next, the supply of food is always very limited.

So furiously hungry are these dogs that it is no uncommon thing for them to eat up the leather harness of the sledges, and at night it is necessary to suspend all such articles out of their reach. Yet, with the carcass of a newly killed reindeer before them, and with the certainty before their eyes of such a meal as they had never enjoyed, and were never likely even to see again, these dogs were generous enough to restrain their appetites, and, instead of gratifying their raging hunger on the dainty banquet within their reach, sat and guarded it for hours, and delivered it untouched to their masters. How many hungry men are there who would have acted so generous a part, and have exercised such trying self-denial? We shall hear more of Barbekark in another portion of this work.

In the *Naturalist's Magazine* there is a remarkable instance of generosity on the part of a sparrow. As a general rule, sparrows are remarkable for their ability in taking care of themselves, and for the manner in which they will seize for themselves the property of others. For example, there are many places where the house-martin used to abound, and is now almost extinct, simply because the sparrows allowed them to build their mud nests, then ejected them, and took possession themselves. Sparrows have also been known to act as the eagle does to the osprey, and the skua-gull to the smaller species, *i. e.*, allow the weaker bird to take all the trouble of capturing prey, and then take it away by violence.

There are, however, exceptions to every rule, and a very honorable one is recorded in the *Naturalist's Magazine*. A lady possessed, among other birds, a canary, whose cage used to hang outside the window. One morning a sparrow perched on the cage, and seemed to hold a sort of conversation with the inmate. Presently he flew away, but shortly returned with a grub, which he dropped into the cage. Every day at the same time the sparrow made his appearance with his accustomed offering, and the canary at last became sufficiently familiar to take his food directly from the sparrow's beak. The lady then put some more cages out of the window, and the sparrow fed all of the inmates, invariably, however, selecting the canary for his first visit, and making the longest stay with that bird.

Now let us pass to generosity in the sense of *Magnanimity*, or unwillingness to resent an injury, though possessing the power to do so. There are few qualities in human nature more noble than the capability of foregoing revenge when the offender is powerless to resist. I suppose that all my readers have heard of the famous answer to a justly offended man, "Would it not be manly to resent such an affront?" "Yes, but it would be godlike to forgive it." Those who are conscious of power are never afraid to forgive; and thus it is that in the daily services of our Church the very first invocation runs thus, "Almighty and most merciful Father." Almighty, and therefore all-merciful. Looking back through history, we shall find that those whose names have lived as the noblest of the human race have been distinguished by that divine quality of mercy which Shakespeare has described in words too familiar for quotation.

Indeed, when we find those beings whom we

call "brute" beasts rising to a moral grandeur which few men can attain, disdaining to avail themselves of the opportunity of vengeance, and even repaying evil with good, it does seem an utter absurdity to say that they are not acting under the inspiration of Him who gave us the celestial maxim, "Love your enemies." By their action they show themselves worthy of life everlasting; and what they deserve they will assuredly receive at the hands of Him who is Justice and Truth.

Consciously or unconsciously, this feeling is acknowledged among mankind. Taking our own nature, for example, prize-fighters are not considered among the most elevated class of society. Yet one of the fundamental rules of the "ring" is, "Don't hit a man when he is down;" and any boxer who demeaned himself by such an act would be at once adjudged to have lost the fight, and would be disqualified from entering the ring for the rest of his life. Striking below the belt is another disqualifying action; and the custom of shaking hands before a fight, and the victor sending round his hat on behalf of his vanquished foe, are customs showing that even in this low stratum of society there is a recognition of the one great principle.

Of the axiom that those who are strongest are least apt to use their strength, a curious example occurred some years ago, when the "ring" was in its palmiest days, and the highest in the land went openly to see a fight as they now go to a horse-race.

A man in the quarrelsome stage of drink came into a public-house, and began to wrangle with those who were already there. At last he took umbrage at one of the guests who was sitting quietly smoking his pipe, and finding that he was not to be drawn into a fight, called him a coward and struck him on the face, drawing blood. The man merely wiped his face, and went on with his pipe. One of the guests exclaimed, "How can you stand this, Tom Cribb?" At the sound of the dreaded name, the assailant dashed out of the room, and was not seen again. Cribb could afford to take an insult from a man whom every one present knew he could have killed at a single blow.

As with man, so it is with the lower animals; and there are many instances on record where the strong have disdained to make reprisals on the weak, no matter what the offense might be.

I knew two dogs in whom the "quality of mercy" was strongly developed. One was an enormous animal called "Lupo," because he looked just like a white wolf, except that he was very much larger. Handsome as he was, his enormous size made him very inconvenient in the house, for when he chose to lie on the hearth-rug no one had a chance of coming near the fire. In the same house was a little black-and-tan terrier called "Tiny." Now in cold weather Tiny liked to have a warm couch by the fire; and whenever Lupo had composed himself to sleep, she used to climb upon his body, turn round and round in his long fur as if he were nothing but a door-mat, and also settle down to rest.

The absurdity of the proceeding was crowned by the fact that when she had thus settled herself she would not allow Lupo to move. If he even ventured to stir and disturb her, she would fly savagely at his head, barking and growling viciously; and if he did not at once lie quiet, thought nothing of biting one of his long ears, Lupo submitting as tamely as if he had taken his name from a lamb, rather than a wolf.

Yet Lupo was by no means an animal to be trifled with. He once had a tremendous fight with his master about a bone; and it was not until after he had bitten his antagonist severely in the wrist and arms, and had had a succession of sticks broken over him, that he succumbed. Having done so, he, after the manner of well-bred dogs, gave in completely, and came crawling to his master's feet for forgiveness.

As to dogs in general, Lupo had an objection to them, and, when he accompanied his master's carriage, had generally to be muzzled, lest he should pick up any stray dog, give it a shake, toss the dead body over his shoulder, and trot on as if nothing had happened. The curious point in his temperament was, that if a dog ran away from him that animal was doomed, unless Lupo had a muzzle. But if the dog flew at him he respected that dog, and treated him with perfect forbearance. I have seen as many as three dogs at a time hanging on to him, Lupo trotting on unconcernedly, and not taking the least notice of them, even when they dropped off through weariness of jaw.

There was one dog which had actually fought himself into friendship with Lupo. He was a terrier belonging to a blacksmith who lived about half-way between the station and the house of Lupo's master. For some time the animal used to fly at Lupo twice daily, namely, during the progress to and from the station; Lupo, as usual, respecting him for his courage, but not even attempting to injure him. At last having, like Mrs. Malaprop, begun with a little aversion, the two animals struck up a friendship, the terrier watching for Lupo, gamboling with him until he had reached his journey's end, and then returning home alone.

My bull-dog "Apollo" was equally magnanimous: he would suffer almost any provocation from a dog smaller or not much larger than himself, but never would allow any liberty to be taken by a big dog. Over and over again has he allowed little dogs to bite him without troubling himself to retaliate; but if a big dog ventured upon an insult, that dog had to run.

One day as I was walking to the post-office, with Apollo at my heels, as usual, a remarkably fine black retriever came up and began to growl at him. Apollo only gave him a glance out of the corner of his eye, and trotted on. The retriever came close, and continued to growl; whereupon I cautioned his owner that, if his dog would let Apollo alone, Apollo would have nothing to say to him, but that if the retriever continued his insults I could not answer for the consequences.

The only reply was a disdainful smile, and a contemptuous look at the dog. After some more annoyance Apollo gave a slight growl, and the hairs of his back began to bristle ominously. Again I gave warning, but with the same result. Presently the retriever flew at Apollo, bit him in the ear, and next moment was on his back, with Apollo's grip on his throat. The retriever's master was so startled at the sudden change of affairs that he could not interfere, and in a minute there would have been a dead retriever. Fortunately I had taught Apollo to loosen his hold at the word of command (the hardest lesson I ever had to teach a dog), and I called him off.

In a few moments the fallen animal recovered his breath and his legs, and made off at full speed, yelping with pain and terror, and I saw nothing more of him. As for Apollo, he fell back unconcernedly to his place at my heels, and trotted on as if such a thing as a retriever had never been in existence.

The anecdotes which have been just related show that animals can act magnanimously toward each other. Here, however, is an instance where a dog which would most assuredly have assaulted the man whom he hated, had the latter been in a position to defend himself, did most nobly forego his vengeance when the enemy was completely at his mercy. As is usually the case with the most characteristic dog anecdotes, the event occurred in Scotland:

"The manager of a mill in Fifeshire was very much disliked by the watch-dog, probably because he had acted harshly to the animal. One very dark night he strayed from the path and fell over the dog. Perceiving the mistake which he had made, and that he could not recover himself, he gave himself up as lost, the dog being a very powerful one. The animal, however, was magnanimous enough to spare a helpless enemy, and to lay aside old grievances. Instead of seizing the prostrate man by the throat, the dog only licked his face and indicated his sympathy. Ever afterward the man and the dog were great friends."

CHAPTER IX.

CHEATERY.

Animal Swindlers.—"Barbekark" as a Cheat.—Roguery Detected.—Dogs Shamming Lameness.—Dogs Cheating Each Other.—The Elephant, "Burra Sahib," Hiding a Cake.—Comparison with Humanity.—Golden-crested Wren as a Cheat and Thief.—Two Ravens Uniting to Cheat a Dog out of his Dinner.—Alliance between a Dog and a Raven.—Principle of the Ambuscade.

ALL virtues have their opposite vices; and just as there are animals capable of exercising great self-denial in order to give to others that which belongs to themselves, and even displaying an amount of generosity unsurpassable by any human being, so there are animals which can cheat like accomplished swindlers. Sometimes, as in the following instance, the same animal is capable of both acts.

Here is an anecdote of "Barbekark," the dog which killed the deer, and then gave it up to his master. The narrator is Captain Hall:

"I have before mentioned some particulars of these dogs, and I now relate an anecdote concerning them during our passage across from Greenland.

"One day, in feeding the dogs, I called the whole of them around me, and gave to each in turn a capelin, or small dried fish. To do this fairly, I used to make all the dogs encircle me, until every one had received ten of the capelins apiece.

"Now Barbekark, a very young and shrewd dog, took it into his head that he would play a white man's trick. So every time that he received his fish he would back square out, move a distance of two or three dogs, and force himself in line again, thus receiving double the share of every other dog. But this joke of Barbekark's bespoke too much of the game that many men play upon their fellow-beings, and, as I noticed it, I determined to check his doggish propensities. Still, the cunning and the singular way in which he evidently watched me induced a moment's pause in my intentions.

"Each dog thankfully took his capelin as his turn came round; but Barbekark, finding his share come twice as often as his companions, appeared to shake his tail twice as thankfully as the others. A twinkle in his eyes, as they caught mine, seemed to say, 'Keep dark; these ignorant fellows don't know the game I am playing. I am confoundedly hungry.'

"Seeing my face smiling at his trick, he now commenced making another change, thus getting *three* portions to each of the others' one. This was enough, and it was now time for me to reverse the order of Barbekark's game by playing a trick upon him. Accordingly every time I came to him he got no fish; and although he changed his position three times, yet he got nothing. Then, if ever there were a picture of disappointed plans, of envy at others' fortune, and sorrow at a sad misfortune, it was to be found on that dog's countenance as he watched his companions receiving their allowance. Finding he could not succeed by any change of his position, he withdrew from the circle to where I was, and came to me, crowding his way between my legs, and looked up in my face as if to say, 'I have been a very bad dog; forgive me, and Barbekark will cheat his brother dogs no more. Please, sir, give me my share of capelins.' I went the rounds three times more, and let him have the fish, as he had shown himself so sagacious, and so much like a repentant prodigal dog."

As cheatery requires the use of the intellect, it is evident that the most intellectual animals will be the most accomplished cheats. Dogs, therefore, may be expected to be considerable adepts in cheating, and are often very amusing in their attempts to deceive human beings. Here are one or two more examples of cheating in the dog:

One of my friends had a couple of little toy terrier dogs. As is usually the case in such instances, though very fond of each other, they were horribly jealous with regard to their master, and neither could endure to see the other caressed. It so happened that one of them broke its leg, and was in consequence much petted. Its companion, seeing the attention that was paid to the injured animal, pretended to be lame itself, and came limping to its master, holding up the corresponding leg, and trying to look as if it were in great pain.

The following anecdote is sent me by a friend:

"A Skye terrier of our acquaintance named 'Monte' had at one time a very sore leg, and during his illness he got a great deal of sympathy and petting. Ever since, when he has been in any mischief, he comes running on three legs, holding up the one which was once sore, but is now quite well. In his own way he is quite as arrant an impostor as the well-known begging 'sailor' with one leg tied up to look as if he had lost it."

A curious and rather ludicrous instance of cheatery on the part of the dog was observed by one of my friends.

He has three little black-and-tan terriers, father, mother, and daughter, which are great pets, and consider the house as their own property. Like most pet dogs, they have their favorite spots by way of couches; and as they all three generally take a fancy to the same spot, there is occasionally a difference of opinion and a slight loss of temper. The one pet spot of all is a soft cushion at the head of a sofa. Now the cushion had accommodated easily the father and the mother; but when the daughter came, and in course of time wanted her share of the couch, it was found that the quarters were rather too limited for comfort, especially as the daughter persisted in growing until she reached the size of her parents.

One day the father and daughter had got into the room first, and according to custom made straightway for the cushion, on which they established themselves comfortably, occupying the whole of its surface. Presently the mother came in, and also went to the cushion. She tried to take her place on it, but her husband was too selfish and her daughter too undutiful to move, and in consequence she had to retire.

Presently she went to the farthest corner of the room, and suddenly began to scratch violently, barking, growling, and sniffing as if she were digging out a rat. Up jumped the others, all blazing with excitement, and anxious to have their share of the sport. As soon as they had got their noses well down in the corner, the mother ran to the sofa at full speed, jumped on the cushion, curled herself round, and was happy. However, she was generous in victory, and made room for her husband and daughter as they came back to the sofa crest-fallen and humiliated.

One of my brothers has furnished me with an account of an audacious piece of cheatery practiced by his dog:

"My dog is a white terrier, called 'Sambo,' on account of his color, supposed to be a pure specimen of the 'fox' variety, but perversely exhibiting unmistakable evidences of the existence of the more plebeian 'bull' somewhere in the roll of his ancestry. He is good-tempered and affectionate, and devoted to his master—and to sport, especially to the pursuit of rabbits.

"One fine morning last January I took him out for a couple of hours' rabbiting, to his great joy, but, as I could also see, by his way of constantly coming back to have a look at my face, to his intense puzzlement. An afternoon alone with me was quite natural, and according to custom; but starting at eleven A.M. had always meant a day with the keepers, and where were the keepers? We found no rabbits; but then he was not busy as usual: his head was not sufficiently clear from other matters to look them up with his usual care and perseverance.

"He passed many a likely bush without even a glance of his eye, and I began to fear that he was ill; when suddenly, as luck would have it, we heard several shots in rapid succession, and found ourselves in the midst of a regular rabbiting party. The effect upon Sambo was miraculous; his tail and ears went up, and he sprang at once from a state of low despondency into one of violent activity. A few moments before and he seemed to have made up his mind that the British rabbit was an extinct animal, and his master a great fool for carrying a gun in pursuit of it: of course, as he was under orders, he must look for them, or pretend to do so—but it was awful humbug. Now, to see him rushing all over the place, quartering the ground, with his tail going, and his nose investigating every little tuft, one would have thought there was a rabbit for every square yard.

"Things went on as usual until the time arrived that I had to take my leave and return home. Now not only had we just arrived at a favorable spot in the covert—a fact just as well known to the dogs as to ourselves—but there were unmistakable signs of approaching luncheon. My first call to my dog was, therefore, unheeded: he had suddenly conceived a violent affection for another dog, with whom, by the way, he could never on ordinary occasions agree, and in the interchange of friendly confidences was quite abstracted from the outer world.

"A more imperative summons made him start —a very false move; but he at once compensated for it by facing round sharply in the opposite direction to me, and looking anxiously up the drive instead of down, with his head and ears up, as if he rather expected to see me at the end of it, about half a mile off. However, it would not do, and he was reduced to following me, though he kept to heel with drooping head and tail, and many a wistful look behind.

"We had hardly got well out of the sight of the keepers when he suddenly brightened up, as though he thought life had yet some joys in store for him, trotted on in front, and behaved himself as usual. Suddenly, just a few yards from the exit from the covert, he 'made a point' at a solitary tuft of grass and rushes. I was astonished that a rabbit could be harbored there, as we had but just passed over the very spot with a regular array of dogs and beaters; but Sambo said 'rabbit' as plainly as possible; so in went my cartridges again, and the necessary permission was given.

"To my astonishment, no rabbit appeared; but none the less Mr. Sambo went through all the regulation manœuvres formulated and provided in such cases. He dashed into the tuft, came out the other side as if in full chase, yelped as if he were only just out of biting distance of his prey, and was lost to sight in a moment; and what is more, he returned not. I whistled and called, but no sound could be heard. Suddenly his 'little game' flashed upon me. I went back to the keepers, and there was my friend taking his luncheon affably with one of them—a particular friend. With the utmost respect for his mental resources, I yet thought it necessary to be 'firm' with him, and I do not think he will ever play me that trick again."

One of the most amusing anecdotes of attempted cheatery is narrated of an elephant by Lady Barker:

"When we paid them a visit upon the afternoon of the storm, the huge beasts were taking a bath, or rather giving it to themselves by filling their trunks with water and dashing it over their heads, trumpeting and enjoying themselves immensely. At a little distance the cooks were busy baking the chupatties—a muffin as large as a soup-plate, and nearly as thick—in mud ovens; and the grass-cutters had been down to a 'jhed,' or pond, near, to wash the dust off the large bundles of grass for the elephants' suppers. We talked a little to the mahouts, and one very picturesque old man seemed exceedingly proud of his elephant's superior slyness and cunning, and begged us to stay and see him 'cheat.' So we waited till 'Burra Sahib,' or 'Mr. Large,' had finished his bath and came slowly up to the mahout for his supper.

"The mahout called out to the cook to bring the chupatties, and made us retire behind a tree and watch what Burra Sahib did. As soon as the cook went away, the elephant put up his trunk and broke off a large bough of the tree above him. This they generally do to serve as a brush to keep off flies, so he knew *that* was nothing remarkable. He then looked slyly around him with his bright, little, cunning eyes; and as he could not see his mahout, he thought the coast was clear, and hastily snatched up a chupattie, which he put under the branch on the top of his head. I noticed how carefully he felt with his flexible trunk if any edge was uncovered, and arranged the leaves so as to hide his spoil completely.

"Burra Sahib then raised his voice and bellowed for his supper in loud and discordant tones. The mahout then ran up as if he had been a long distance off, stood in front of him, and commenced handing him the chupatties, counting, as he did so, one, two, three, and so on. The elephant received each in his trunk, and put it gently into his huge mouth, bolting it as though it had been a small pill. Twelve chupatties was the allowance, and he required this sort of food to keep him in good condition. When the mahout came to number eleven muffin he looked about for the twelfth in great dismay, pretending that he could not think what had become of it, and calling for the cook to scold him, searching on the ground, and wondering, in good Hindostanee, where that other chupattie could be. The elephant joined in the search, turning over an empty box which was near, and trumpeting loudly.

"The mahout was delighted to see how much this farce amused me, and at last he turned suddenly to the elephant, who was still hunting eagerly for the missing chupattie, and reviled him as a thief and a 'big owl,' adding all sorts of epithets, and desiring him to kneel down, which Burra Sahib did very reluctantly. The mahout then scrambled up on his head, snatched off the branch, and flung down the chupattie, belaboring the elephant well with the bough which had served to conceal it. It seems that the trick had been played successfully many times before Burra Sahib was found out, and the poor cook used to get into trouble, and be accused of keeping the missing chupattie for his own private consumption."

A servant belonging to one of my friends acted just like this elephant. She had broken a valuable China vase, and in order to hide the evidences of her delinquency she broke up the fragments very small and buried them. When the vase was missed, she protested that she knew nothing about it. She knew that such a vase was somewhere in the house, but had not the least idea where it could be; and for three whole days she went over the house with her mistress, hunting in every cupboard and on every shelf for the article which she had herself buried in the garden.

Birds can be capable of cheating, not only each other, but other animals. Even the pretty little golden-crested wren has been detected in deliberate theft and deception.

A gentleman was watching a chaffinch building its beautiful nest, and soon found that he was not the only spectator. At a distance was perched a golden-crested wren, which watched the proceedings carefully. As soon as the chaffinch went off to fetch more materials, the gold-crest cunningly stole round in an opposite direction, and carried off the newly brought hairs, etc., for its own nest. This went on for some time, until at last the aggrieved chaffinch found out the robbery, and chased the gold-crest so fiercely that it did not attempt to renew the theft.

This story is told by Mr. W. Thompson, in his "Natural History of Ireland." He also states that this kind of robbery is not at all uncommon with the gold-crest. Its nest is made of the same material as that of the chaffinch, and so it is accustomed to avail itself of the labors of that bird in order to lighten its own toil.

The celebrated arctic voyager, the late Sir L. M'Clintock, mentions a curious instance of stealing on the part of the raven. When they were in Mercer Bay, a pair of ravens, probably male and female, used to hang about the ship, and pick up any refuse food that might be lying about. At a certain hour of the day the men were accustomed to wash out their mess-tins, the rejected contents of which were regarded by the ship's dog as his proper perquisites. The ravens, however, held a different opinion, and by force of superior intellect almost always contrived to gain their own ends.

As soon as the tins were emptied, and the dog ready for his meal, the ravens set to work to cheat him out of his food. They assaulted him from the front, keeping him from his food by perpetual annoyance, and at last induced him to make a charge at them. Of course, after the manner of ravens, they contrived to flap their way just out of his reach. This process was repeated until they had inveigled him to a considerable distance, when they took to wing, and being able to fly faster than the dog could run, managed to secure a good meal before he could reach them.

It is evident that they must have concerted this plan of action between them; so that we see in this ruse an example of reason and the communication of ideas by means of language. My readers may perhaps remember the story of the two dogs who used to hunt the hare in concert, the one starting the hare and driving it toward the spot where the accomplice lay hidden. I knew of an instance where a somewhat similar arrangement was made; only in this case the two contracting parties, instead of being two dogs, were a dog and a raven, the latter making use of its wings in driving the prey out of the heather into the open ground.

Many instances of such alliances are known, and in all of them there is the curious fact that two animals can arrange a mode of cheating a third. In fact they employ one of the principal stratagems in the art of war, *i. e.*, the ambuscade, or inducing the enemy to believe that danger is imminent in one direction, whereas it really lies in the opposite and unsuspected direction. No one would say that a general who contrived to draw the enemy into an ambuscade acted by instinct: the act would be accepted as a proof of reasoning powers surpassing those of his adversary. And if this be the case with the man, why not with the dog, when the deception is carried out by precisely the same train of reasoning?

CHAPTER X.

HUMOR.

Practical Joking the Lowest Kind of Humor.—Torture the Humor of the Savage.—Spinning Cockchafers. —Making a Boy "Jump like a Dog."—Humor in Birds.—The "Chukor" Partridge and the "Punkah-wallah."—Humor in the Buzzard.—The Kestrel's Idea of Humor.—The Humorous Heron.—"Making Believe" in Children and Animals.—Swallows Mobbing a Kestrel.—The Same Birds Mobbing a Hare.—Swallows Mobbing a Cat.—Spar-Ousels Mobbing a Cat.—Jackdaws Doing the Same.—Ring-Doves Mobbing a Dog.—Monkeys and Crocodiles.—A Cock Tantalizing the Hens with Food, and Eating it Himself. —Sense of Humor in the Parrot.—A Cat Deceived by a Parrot.—The Peacock and the Poultry.—Humor in the Emeu.—Humor in the Mooruk.—A Dog and his Practical Jokes.—A Horse Playing Practical Jokes on a Boy.—Mr. G.'s Pony.—Horses Chasing a Pig.—Animals Joining in Children's Sports.—A Dog Playing at "Touch."—A Pony and a Cat Playing "Hide-and-Seek" with the Children.—"Peter," the Field-Mouse, Playing the Same Game.—A Mischievous Canary.—The Bullfinch and the Work-box.—Practical Jokes Played by "Ungka," the Siamese Ape.—Sense of Humor in the Next World.

AMONG other traits of character which are common to man and beast is the sense of humor.

This is developed in various ways. Mostly it assumes the form of teasing or annoying others, and deriving amusement from their discomfort. This is the lowest form of humor, and is popularly known among ourselves as practical joking. Sometimes, both with man and beast, it takes the form of bodily torture, the struggles of the victim being highly amusing to the torturer. Civilized man has now learned to consider the infliction of pain upon another as any thing but an amusement, and would sooner suffer the agony than inflict it upon a fellow-creature. But to the savage there is no entertainment so fascinating as the infliction of bodily pain upon a human being.

Take, for example, the North American Indian tribes, among whom the torture is a solemn usage of war, which every warrior expects for himself if captured, and is certain to inflict upon any prisoner whom he may happen to take. The ingenuity with which the savage wrings every nerve of the human frame, and kills his victim by sheer pain, is absolutely fiendish; and yet the whole tribe assemble around the stake, and gloat upon the agonies which are being endured by a fellow-creature. Similarly the African savage tortures either man or woman who is accused of witchcraft, employing means which are too horrible to be mentioned.

Yet even in these cases the cruelty seems to be in a great degree owing to obtuseness of perception; and the savage who ties his prisoner to a stake, and perforates all the sensitive parts of his body with burning pine-splinters, acts very much like a child who amuses itself by catching flies, pulling off their wings and legs, and watching their unavailing efforts to escape. I do not know whether it is the case now or not, but some twenty years ago I saw cockchafers publicly sold in Paris for children to torture to death; the amusement being to run a hooked pin through its tail, tie a thread to it, and see the poor insect spin in the air. After it was too enfeebled to spread its wings, it was slowly dismembered, the child being greatly amused at its endeavors to crawl, as leg after leg was pulled off. I rescued many of these wretched insects from the thoughtlessly cruel children, and released them from their sufferings by instantaneous death.

In Italy a similar custom prevails, though in a more cruel form, the creatures which are tortured by way of sport being more capable of suffering pain than are insects. Birds are employed for the amusement of children, just as are the cockchafers in France. A string is tied to the leg, and the unfortunate bird, after its powers of flight are exhausted, is generally plucked alive and dismembered.

It is not done from any idea of cruelty, but from sheer incapacity to understand that a bird or a beast can be a fellow-creature. The Italians are notorious for their cruel treatment of animals; and if remonstrance be made, they are quite astonished, and reply, "Non è Cristiano" (It is not a Christian).

Not that we in England have very much

to boast of on this score. The Puritans did a good work when they abolished bear-baiting, even though, as Macaulay says, they did so not because it gave pain to the bear, but because it gave pleasure to the spectators. But up to the present day there is a latent hankering after similar scenes, even though they are now contrary to law; and dog-fighting, cock-fighting, badger-drawing, and rat-killing are still practiced in secret, though they can not be carried on in public.

Mr. W. Reade, in his work on "Savage Africa," mentions a case in which a woman and her son, a young boy, were put to death on a suspicion of witchcraft. The woman was drowned, and the boy burned alive; sundry packets of gunpowder being tied to his legs, which made him "jump like a dog"—thus causing great amusement to his torturers. Mr. Reade remonstrated with them upon their cruelty, but they could not be made to see that there was any more cruelty in the fate of the son than in that of the mother. The narrator was astounded at the very notion. "Burning more bad! No, Mr. Reade, burning and drowning all the same." The cruelty was not intentional—it was simply want of understanding. To see the boy "jump like a dog" was highly amusing to the spectators, and they never troubled themselves about the fact that the ludicrous contortions were caused by terrible pain. Savages are in many points nothing but children, and they act after childish manners, but with the powers of men for evil.

For example, the poor little boy of seven years of age who was afterward so cruelly burned alive was subjected at the hands of his captors to a species of humor which was vastly entertaining to them. "On the ground crouched the child, the marks of a severe wound visible on his arm, and his wrists bound together by a piece of withy. I shall never forget that child's face. It wore that expression of passive endurance which is one of the traditional characteristics of the savage. While I was there, one of the men held an axe below his eyes: it was the brute's idea of humor."

In a similar manner, the sense of humor is mostly developed in the lower animals by causing pain or annoyance to some other creature, and the animal acts in precisely the same manner as a savage or a child.

We will just take a few cases of humor as exhibited by birds.

As might be expected from the character of the birds, sparrows will gratify their feelings of dislike by uniting together for the purpose of mobbing some creature to which they have an objection. There is a short account in Hardwicke's *Science Gossip* for December, 1872, of a number of sparrows mobbing a cat.

The cat evidently intended to make a meal on one of the birds, but was greatly mistaken; for the sparrows dashed at him so fiercely that he soon turned tail and ran into the house, one of the sparrows actually pursuing him into the house. Poor Tommy ran up-stairs, and was found crouching in terror under one of the beds. This happened in London, where, by the way, sparrows are much less numerous than they used to be: this, I believe, is chiefly if not entirely due to the staff of street-cleaning boys, who remove the substances from which the sparrows used to derive the greater part of their subsistence.

An account of a somewhat similar adventure is given in the *Dumfries and Galloway Standard*.

In the year 1856 a number of "rooks" were in the habit of assembling on a house, and it was thought that they had nests there. One day a cat came prowling over the roof, to the great discomfiture of the rooks, who assembled on the roof of a neighboring house and held a consultation. This being over, they proceeded systematically to attack the foe, dashing at her in groups of three or four, flapping their wings in her very face, and screaming dismally.

As for the cat, though a young one, she was not in the least dismayed, her courage rising to the occasion.

"It then," writes a spectator of the scene, "looked the very image of defiance; and a more graceful figure of a cat we never saw when, in fighting attitude, it strained its head and struck out its dexter paw. The cat frequently changed its position, with the view, we suppose, of doing its best to bring itself into closer quarters with the crows; but in vain. They kept up a shower of abusive language, and occasionally almost grazed the head of Grimalkin with their feathers, but they never ventured going within her reach. Puss mewed impatiently at times, as much as to say, 'Oh that I had wings for a few minutes, and then I would put an end to your noise and bluster.'

"This skirmish between this valorous cat and the crows lasted fully half an hour, and ended in a sort of drawn battle. The cat would have kept the castle long enough, in despite of them; but it could neither get hold of them nor their nests, so it at length quietly descended."

In his "Natural History of Ireland," Mr. Thompson records a case where that rather rare bird, the ring-ousel, mobbed and drove away a dog.

Mr. Thompson was shooting in the Crow Glen, accompanied by his pointer, which was some yards in advance. Suddenly the dog was attacked by two male ring-ousels, which dashed at its head, accompanying each stroke with loud shrieks. They were incited to this action by a female, which, after setting them at the dog, retired to a distance and looked on at the fight. The dog was so alarmed by the attack of the birds that he was obliged to retreat to his master. The birds were so determined in their onset that they even attacked Mr. Thompson and two of his friends, who were accompanying him.

Had these birds been male and female, it might have been thought that they were defending their young, or trying to decoy the dog from their nest; but they were both males, and in their adult plumage. The action lasted for more than a quarter of an hour.

All those who have watched the habits of animals must have remarked how widely spread among them is this species of humor—namely, annoying and insulting a stronger being than themselves whenever they think that they can do so with impunity. And so strong is the impulse to gratify their sense of humor that they do not hesitate to do so at the risk of their lives. M. Mouhot, in his work on "Indo-China and Cambodia," mentions that he has often witnessed very amusing scenes between the monkey and the crocodile.

The latter animal is lying half asleep on the bank, and is espied by the monkeys. They seem to consult together, approach, draw back, and at last proceed to overt acts of annoyance. If a monkey can find a convenient branch, he goes along it, swings himself down, hangs by a hand or a foot, slaps the crocodile on the nose, and instantly scrambles up the branch out of the reptile's reach. Sometimes, when no branch is sufficiently near, several monkeys will hang to each other so as to make a chain, and, swinging backward and forward over the crocodile's head, the lowermost monkey will torment the reptile to his heart's content. The cream of the joke is when the crocodile is at last so irritated that it opens its enormous jaws, makes a vicious snap at the monkey, and just misses him. Whenever this happy event occurs, there are screams and chatterings of exultation from the monkeys, and a vast number of joyful gambols executed among the branches.

Of course, according to the old proverb, the pitcher may go too often to the well; and it does occasionally happen that the monkey does not escape in time, and is ingulfed in the crocodile's jaws. Whereupon the cries of exultation are changed into groans and shrieks of terror, and the whole assembly make off as fast as they can. But experience does not teach them discretion, and in two or three days they will be at the same game again.

In Mr. T. C. Jerdon's "Birds of India" there is an amusing notice of the habits of the Chukor partridge (*Caccabis Chukor*) when domesticated. It is very tame and familiar, and sometimes becomes rather a nuisance on account of its habit of playing tricks on people. It has a special facility in discovering the most vulnerable spot, and inflicts sly pecks at the bare feet of the native servants as they move through the house. Its great amusement, however, is to find the man who pulls the punkah half asleep, as is customary with these men as they rock backward and forward at their monotonous task. The little bird pecks his legs so fiercely and actively that he is quite unable to drive it away and go on with his work, and he is at last obliged to call for some one to rid him of his tormentor.

A somewhat similar custom is related by Mr. Thompson, in his "Natural History of Ireland," the bird in this case being a tame buzzard. It had a way of flying after strangers, and knocking their hats over their eyes with a blow of its wing; and it was so quick about it that, even when forewarned, its victim had some difficulty in evading the blow. The same bird took an objection to the bow of its master's shoe-ties, and used to fly at his feet and suddenly untie the strings.

I am personally acquainted with a heron in which this form of humor is largely developed.

The bird was in one sense tame, for it was allowed to run loose in a garden, and was on the most affectionate terms of friendship with one of the men employed in its owner's warehouse. It is really beautiful to see the welcome which the bird gives to the man, and to hear her low, loving gabble as she rubs her head against him, or takes his hand gently in her beak. He has taught her several tricks, as, for example, to take off his hat at the word of command. She is a beautiful creature; and if the sense of humor were not quite so strong, she would be an admirable bird. Unfortunately, she has an unappeasable relish for practical jokes, especially against human beings, looking quite soft and gentle until they are within reach, and then driving her long, sharp beak at them with the rapidity of a serpent's stroke.

On one occasion her sense of humor was developed to such an extent as to cost the loss of her liberty.

The garden in which she lives is also inhabited by a great number of aquatic birds, principally gulls and ducks, and they have a way of laying their eggs in different parts of the garden. One day a learned and respected neighbor went into the garden, and, seeing some ducks' eggs on the ground, stooped down to examine them. As he was thus engaged, the heron stole up softly behind him, and delivered so tremendous a blow that she fairly knocked him on his face.

The unfortunate gentleman, knowing that the heron has an unpleasant way of pecking at eyes, crouched as closely to the ground as he could, sheltering his eyes with his arms, and calling for help as loudly as he could in such a position. The heron, enjoying the joke immensely, mounted on his back, and triumphantly maintained her post there until assistance arrived, and she was driven off.

Since that exploit she has not been allowed to run loose, but has been confined in a roomy cage, in which she can run about. Even under these circumstances she delights in enticing people to come near the cage, and then darting her beak at them between the bars—a joke which she has several times played at my expense. The distance to which she can project her beak is quite marvelous, and it is no difficult matter for the bird to decoy too confiding persons within her reach.

Humor, indeed, seems to be a special characteristic of the hawk tribe. I knew a tame sparrow-hawk which was always trying to circumvent a magpie belonging to the same house; and the extraordinary ingenuity which these birds showed in playing practical jokes on each other could not have been surpassed by human beings. In Hardwicke's *Science Gossip* of March, 1871, there is an account of a tame kestrel which showed a similar sense of humor:

"Insects of all sizes and kinds were summarily devoured, and I have more than once captured wingless females and imperfectly formed moths unable to fly, by finding her dancing round them in their endeavors to escape, and with a gentle nibble giving them a hint to run faster. When the poor insects were too maimed or exhausted to crawl farther, the sport being ended, they were eaten without further delay. Indeed, she appears quite indignant with spiders, because, instead of hurrying off, they lie down and curl themselves up."

Here we have an instance of a bird dealing with an insect just as the French children dealt with the cockchafers, neither bird nor children having the least idea that the struggles which amused them so much were the result of pain.

The same bird, if she could find neither mouse nor insect, would pounce on a piece of brick or stone, and carry it off in her claws, making believe that it was prey of some sort. She carried on this pretense to such an extent that she would resent any interference, and would fight for her piece of brick as fiercely as if it had been a mouse, of the delicacy of which she was most fond. No child could have "pretended" with more abandonment than she did, and the bird which "made believe" that the piece of brick was a mouse, and the child who "makes believe" that a piece of stick is a baby, are for the time, and on that point, precisely on a level.

The following account of bird humor, as displayed in practical joking, is taken from Hardwicke's *Science Gossip* of March, 1872:

"I have imbibed many of the tastes of Gilbert White; but that which engrosses me most, and which I may call my hobby, is the natural history of the swallow tribe.

"I have read that swallows will 'mob' and put to flight a kestrel-hawk. This I was rather skeptical of until lately, when my doubts were removed by that most convincing of proofs, ocular demonstration. I had gone to see an old castle in the neighborhood, which was built on the only hill for miles around, and was therefore tolerably certain to be the haunt of a pair or two of hawks. I accordingly kept my eyes open, in the expectation of seeing one, and I was soon rewarded by the appearance on the brow of the hill of a bird which, by its graceful form and the hovering motion of its wings, I knew to be a kestrel.

"His active little enemies, the swallows, a flock of which were disporting themselves close by, had been as quick to see him as I. These at once advanced to meet the intruder, and, with the utmost audacity, brushed past him in all directions, one from one quarter and one from another, each wheeling after it had swept by, and returning to the charge, while the hawk made futile dashes now and again, but was always too late to do any damage to his nimble little opponents.

"At last, tired of waging an unequal war, and obliged to own himself conquered, he beat a hasty retreat. He was not, however, allowed to get off so easily, but was followed up by his victorious foes; and the apparent mystery of such little birds proving more than a match for such a formidable-looking antagonist, armed literally *cap-*

a-pie, as he was, was quite cleared up; for as he made off, evidently at his best speed, the swallows, with the utmost ease, when left at an apparently hopeless distance behind, fetched him up, then passed him (in what appeared to me most dangerous proximity), wheeled round, met him on their return journey, and then, taking another sharp turn to the right about, repassed him, and continued repeating these manœuvres a dozen times or more.

"The solution of the mystery lay in their extraordinary power of flight. The way in which the swallows made straight for him, apparently bent on a personal encounter, and then, when the kestrel was reckoning on clutching them in his talons, gliding away at a tangent, was, though no doubt tantalizing to the hawk, none the less amusing and interesting to me.

"To crown all, when the others had left off the chase, presumably not thinking it worth their while to pursue any farther, it was curious to watch one solitary individual carry it on alone with such seemingly unrelenting vigor that he seemed actuated by feelings of the direst revenge. However that might be, the swallow certainly effectually prevented the discomfited foe from pausing in his enforced retreat. I watched them until pursuers and pursued both vanished from my sight. I dare say the little swallow continued the pursuit until he had wearied and exhausted the hawk.

"On another occasion I witnessed a little incident which has, to the best of my knowledge, the merit of novelty; and so I hope you will excuse my telling it. I saw a hare running across a large park by the way-side, and was looking about to see what had started it, but could not imagine what it could be, as neither man nor dog was in sight. It started again (for it had stopped and sat in a listening attitude), and then I saw that the disturbers were a flight of swallows, who were following it up like a pack of hounds; now one and now another skimming past the hare's ears along the ground, while the poor timid creature was putting its best leg foremost; but all to no purpose, for its relentless tormentors seemed to take pleasure in its fright, and to enjoy the sport of teasing it.

"I followed the little group until an undulation of the park hid it from my view, and was greatly surprised to see the dexterity with which the swallows calculated their distance so as to impress the hare with the idea they were flying straight at her, and yet, when they were on the point of dashing against her, took a sharp turn, and swept off in a curve, to renew the attack again the next moment.

"I will close my epistle with an anecdote related by the Rev. Philip Skelton, as having come under his own observation, which seems to be appropriate, and which, I believe, will be new to most if not all your readers. I give it in his own words:

"'I have entertained a great affection and some degree of esteem for swallows, ever since I saw a remarkable instance of their sense and humor played off upon a cat which had, upon a very fine day, rested herself upon the top of a gate-post, as if in contemplation, when ten or a dozen swallows, knowing her to be an enemy, took it into their heads to tantalize her in a manner which showed a high degree not only of good sense but of humor. One of these birds, coming from behind her, flew close by her ear, and she made a snap at it with her paw, but it was too late. Another swallow, in five or six seconds, did the same, and she made the same unsuccessful attempt to catch it; this was followed by a third, and so on to the number just mentioned; and every one as it passed seemed to set up a laugh at the disappointed enemy very like the laugh of a young child when tickled. The whole number, following one another at the distance of about three yards, formed a regular circle in the air, and played it off like a wheel at her ears for near an hour, not seemingly at all alarmed at me, who stood within six or seven yards of the post. I enjoyed this sport as well as the pretty birds, till the cat, tired out with disappointment, quitted the gate-post, as much huffed, I believe, as I had been diverted.'"

The habit of "mobbing" seems to be inherent in animal nature generally, and is even found in fishes, as may be seen by the following anecdote. It is related by Captain Crow, from personal observation:

"One morning during a calm, when near the Hebrides, all hands were called up at 3 A.M. to witness a battle between several of the fish called threshers, a fox-shark, and some sword-fish on the one side, and an enormous whale on the other. It was in the middle of summer, and the weather being clear, and the fish close to the vessel, we had a fine opportunity of witnessing the contest.

"As soon as the whale's back appeared above the water, the threshers, springing several yards into the air, descended with great violence upon the object of their rancor, and inflicted upon him the most severe slaps with their long tails, the sound of which resembled the report of muskets fired at a distance.

"The sword-fish, in their turn, attacked the

distressed whale, striking it from below; and thus, beset on all sides, and wounded, where the poor creature appeared the water around him was dyed with blood.

"In this manner they continued tormenting and wounding him for many hours, until we lost sight of him, and I have no doubt that in the end they completed his destruction."

It is worthy of notice that, in this case, a temporary alliance was formed between fishes belonging to different families. The sharks and swordfishes have but little in common, and yet they united in order to attack the whale, which could not have done any harm to either of them. It is evident, therefore, that fishes must be able to communicate ideas to each other, and to act upon those ideas. In other words, they possess a language which is intelligible to fishes in general, and not restricted to any one species. It is absolutely inaudible and unintelligible to us, but that it exists is an absolute certainty.

A still more curious alliance was mentioned to me by Captain Scott, R.N.—namely, a joint attack upon a whale by the grampus and swordfish, i. e., an alliance between a mammal and a fish against a mammal.

Birds seem to be great adepts in the art of tormenting, and this talent accordingly shows itself where least expected. As a rule, domestic poultry are remarkable for the generosity with which the master-bird treats his inferiors: he will scratch the ground, unearth some food, and then, instead of eating it himself, will call some of his favorites to him, and give to them the delicacy for which he had labored. But I knew of one case—a solitary one, I hope—where the cock scratched as usual, called his wives, and, when they had assembled round him, ate the morsel himself. It was just like the old school practical joke. Old boy to new boy, holding out an apple: "Do you like apples?" New boy to old boy: "Yes." Old boy to new boy: "Then see me eat one."

Parrots are possessed of a very strong sense of humor, and are much given to practical joking, after the ways of mankind. My own parrot had a bad habit of whistling for the dog, and then enjoying the animal's discomfiture; and there have been many parrots who would even play practical jokes on human beings. Dogs and cats, however, seem to be the principal victims of the parrot's sense of humor.

I know a case where a parrot is allowed to go about the house as it likes. In that house there is also a cat, with which Polly is pleased to amuse herself. One day when the cat was lying asleep on the rug, the parrot began to mew and scream just like young kittens when they are hurt. Up jumped the cat, and rushed in frantic haste to her beloved offspring, and was very much astonished to find them all safe and comfortably asleep. She then returned to the rug; but as soon as she had curled herself up and settled herself comfortably, the parrot recommenced her mewings and cries, and in this way contrived to dupe the cat three times.

Every one who has watched the habits of peacocks knows the peculiar rustling sound which they can produce by shaking the feathers of their train. One of these birds, which inhabited a large yard in common with other poultry, was pleased to take umbrage at the chickens, and amused himself by driving them about, and not allowing them to eat their food. His crowning joke was to drive them all into a corner, spread his train, and rustle the feathers over their heads so as to frighten them.

All birds of the gallinaceous order are horribly alarmed at any thing that appears above them, probably owing to their instinct which teaches them to beware of a bird of prey. Sportsmen who have found the birds become wild and wary toward the end of the summer are well aware of this fact, and by flying a common paper kite are enabled to come quite close to the birds, which mistake the paper kite for a bird of prey, and crouch closely to the ground as long as it is above them. The peacock was therefore playing on this instinctive sense of fear when it spread its train over the chickens.

In his "Gatherings of a Naturalist in Australia," Mr. Bennett mentions an instance of humor in an emu. A pair of these birds lived at Sydney, and were so tame that they walked about among the people who came to listen to the band. One day some persons were present who did not know the birds, and, being afraid of them, ran away. Whereupon the emus, enjoying the joke, gave chase after one of the fugitives, and took off his hat.

The same author gives a description of the beautiful species of cassowary called the mooruk. He kept a pair of them in a yard with his poultry, among which was a very consequential bantam cock. Every now and then the mooruk would take a fancy for chasing the bantam all over the yard, and endeavoring to trample him underfoot.

Here are two accounts of a similar mode of practical joking carried on by a dog, which I knew personally, and a horse, both belonging to the same lady:

"We have a little Pomeranian dog, one of whose principal amusements consists in persecuting any fowls which may invade the precincts of his garden, though he never meddles with them when they keep to their own territories. His favorite mode of torture consists in running down the unfortunate fowl, rolling it over upon its back, and then running round and round it. This conduct the dog repeats as often as the poor victim regains its feet. Should the fowl happen to be a large Cochin or Malay, the frantic agitations of its elevated legs are most ludicrous."

Horses, when kindly treated, are very fond of practical joking, from sheer exuberance of spirits. Ignorant grooms very often are unable to understand that playfulness is not vice, and when they are brought in contact with a high-spirited, playful animal, consider it to be a vicious one, and treat it with brutal violence, thus ruining the temper of the animal. Here are some examples of practical joking in horses:

"One of our carriage-horses, 'Charley,' although by no means vicious, was a saucy creature. We had much difficulty in securing him, as he could slip or untie his halter, take down the bar, and open the stable door. One day the groom forgot the necessary precaution of locking the door. Out into the yard walked Charley, where he found the coachman's little boy. The animal did not attempt to hurt the child, but (with that feeling which causes great boys to find amusement in teasing younger ones) drove him into a corner, and seeing that the little fellow was frightened, kept him there by shaking his head at him whenever he attempted to escape. I happened to be the first person who discovered them, and, although but a child myself, went to the rescue.

"I knew the animal's funny tricks, and he knew that I was not afraid of him; therefore he allowed me to lead him back to the stable, only giving a parting shake of the head to his late prisoner. Although so fond of liberty himself, he would thus imprison dogs, cats, or fowls whenever an opportunity offered."

One of my friends, when a boy, had a Shetland pony, whose idea of humor consisted in throwing every one who got on his back; and the variety of means which he could employ showed a wonderful readiness and fertility of invention. Having heard the owner of the pony tell a few anecdotes of his former favorite, I asked for further details, and received from the old coachman of the family the following account, which I print exactly as it was written:

"In the year 1841, 2, 3, The P—— Fox Hounds was kept at K——, J. G., Esq., master of them. There was three young Gentlemen, sons of Mr. G. They had each Poneys for hunting. Mr. F. was the eldest, then Mr. C. and Mr. A. Mr. F. twelve years of age, Mr. C. ten, and Mr. A. nine. The Poneys was kept rough, never in a Stable; they ran out in the Park summer and winter, had a shed to go into at night; they got a little Corn and Hay in winter, that was all the Grooming they got. One of them named Tom tit was rather a rum one to ride; he was about ten hands High, a dark Bay with Black Points, Carraed very little flesh, more like a roe deer than a Poney; his rider was Mr. C. At that time his weight would be about 6 Stone. None Could ride him but him self. I remember of 5 of the Stable Lads trying to ride him in the Park amongest the rest I was one. No sooner did we get on to him then he Pitched every one of us over his head. Of Course we had no Saddle.

"I have seen Mr. C. get on to him in the Stableyard, and the first thing he would try to do was to Pitch him over his head; having failed in that, he would try to rub him off against a wall or house, thinking he had got his leg betwixt him and wall; but Mr. C. was too wide awake for him, the moment he saw what he was up too he put his leg up on his neck, then having faild there, he maid for the Coachhouse wich was verry narrow, Just room enough to let a Man in along side of Carriage, he would get in there trying first the wall and then against the Carriage; he had not room enough to turn to get out, so that some of us would lift him and his rider out without any Damage being done; then having faild in all these atemps, he ran off Past the Mansion house; there is a burn runs Close Past the house, a Bridge over it, and then a gate about 4 feet High, and wich he maid an atemp to Jump, he got his fore legs over the tope bar up as far as the knees and he was fixed he Could nether get one way nor another, he was Standing on his hind feet almost as Straight as a man and his rider Still in the Saddle. I run up to him and said, What are you doing there? His rider said, I know what, he wants me off some way or other, so must Just get down when he had the boldness to get up, I will not Come off: so I lifted the two off the Gate. I have seen him when in a run with the Hounds go through a hole in a hedge you would think not large enough to let one of the Hounds. He was very Seldom thrown out of a run, he Could gallop like a race horse; very Good for soft ground, being so light. I remember of the going a-missing all in a sudden so they were both lost. Mr. G. called out, C., were are you? The answer sounded near where were Standing I hear they were both lying in a ditch up to the neck in water, and Poor C. and Poney had to go home very much aganst there will."

The same pony was afterward sold and taken away. However, when he was some five or six miles away, he had recourse to his old tricks: he sent his rider over his head, galloped off, and jumped several walls, swam the river Earn, and presented himself at his old quarters.

The "Mr. C." mentioned by the coachman tells me that the perpetual struggle for mastery was nothing but pure fun on the part of the pony; but that if he had once dismounted, even when in such absurd positions as those which the coachman so well describes, the pony would have been master ever afterward.

Last summer I was witness to a scene showing that the horse possesses a strong sense of humor. I was walking through Barfreyston, a village near Dover, and saw over the rather high wall of a farm-yard a couple of horses careering about madly. The wall was so high that only their heads could be seen, and occasionally a whisk of their tails. Finding an aperture through which I could look without being seen, I found that the horses were amusing themselves by chasing a pig, hunting it round and round the yard, driving it into corners, and occasionally flinging their heels into the air with delight.

They scarcely gave that wretched pig any rest. Sometimes, when tired with their exertions, they would lie still for a few minutes, and the pig would get away as far as possible from his tormentors. But no sooner did the poor animal settle down to a cabbage leaf than the horses would be at him again, driving him about, and putting him in such a state of perturbation by chasing him from different directions that he had not the least idea where to run so as to escape his tormentors. The horses, in fact, were acting just as two school-boys might be expected to do if a pig's adverse fate had delivered it in their hands.

Many of the lower animals not only show their playfulness in such tricks as those which have been mentioned, but are able to appreciate and take part in the games played by children. When I was a boy I knew a little dog, a King Charles spaniel, which was an accomplished player at the well-known game called tag, or touch. The little animal displayed quite as much enthusiasm as any of the human players, and would dart away from the boy who happened to be "touch" with an anxiety that almost appeared to be terror. Of course to touch the dog was an impossibility; but he was a generous little creature, with a strong sense of justice, and so, when he thought that his turn ought to come, he stood still and waited quietly to be touched. His mode of touching his playfellows was always by grasping the end of their trousers in his teeth; and as it was impossible for the boy to stop when so seized in full course, the dog often got jerked along the ground for some little distance.

A lady told me lately that, when a girl, she had a pony which would play hide-and-seek with the children. Hiding was necessarily only a pretense on the part of the pony; but the animal would go to some corner, hide its head, and make believe that it was entirely concealed.

Hide-and-seek seems to be a game which can be learned and enjoyed by many animals. One of my correspondents has sent me an account of a favorite cat which was an adept in the game. She was a white cat with yellow eyes, and went by the name of "Daisy." She was given, when quite a little kitten, to her mistress, who was then a young child, and the two became inseparable companions, joining in their sports, one of which was hide-and-seek.

The little mistress used to hide, and the kitten to search for her, invariably discovering her lurking-place.

One of the most curious points in this animal was that when she became a cat and had a kitten of her own, she taught her young one the game which she had learned from her mistress, importing into the game an element which I have over and over again seen in the same game when played by children. The kitten went and hid itself, or rather pretended to hide, and the mother went in search of it. She would pretend not to see the kitten, and pass close to its hiding-place. Then, as if startled, she would spring back, the kitten would jump out at her, and the two would rush about in high glee.

The reader can compare with this story the anecdote of "Pop" and the hidden key, which will be found in another part of this work.

A somewhat similar anecdote is told in the *Zoologist*, page 9430, of a short-tailed field-mouse, which had been tamed. It was found so covered with ticks that it could hardly crawl. It was picked up, cleared from the vermin, and placed in a box. It was so grateful for the relief that it did not try to escape, and on the very first day took food from the hand of its benefactor.

"Little 'Peter,' as he was named, soon learned to come when called, and was let out of the box every day to play about the room. Strange to say, he showed a decided appreciation of fun, a favorite amusement being to hide himself in a basin of corn which was kept for his benefit. In this he would bury himself, refusing to answer to his name, and evidently expecting to be looked for. If my friend took no notice of him, Peter's slender stock of patience soon became exhausted; first a shrill squeak was heard, then the corn flew up in showers, and at last up came Peter's little round head to the surface."

This interesting little animal died from feasting too freely on a pear which had been injudiciously given to it by one of the servants.

Dr. Bennett, in his "Gatherings of a Naturalist in Australia," mentions that a couple of young duck-bills in his possession used to play at hide-and-seek behind the furniture of the room. One would hide itself and then give a squeak, when the other would hunt for it and at last find it.

The reader will remember that a kestrel possessed the same powers of "making believe," pretending that a piece of brick was a mouse, and fighting fiercely if any one offered to take it away.

Not even the huge and unwieldy whales are exempt from the sense of humor as displayed by playfulness. In Bennett's "Whaling Voyage" there is a short and graphic description of this trait of character in the spermaceti-whale, or cachalot, as it is often called:

"A large party of cachalots gamboling on the surface of the ocean is one of the most curious and imposing spectacles which a whaling voyage affords; the huge size and uncouth agility of the monsters exhibiting a strange combination of the grand and ridiculous.

"On such occasions it is not unusual to observe a whale of the largest size leap from the water with the activity of a salmon, display the entire bulk of its gigantic frame suspended at the height of several feet in the air, and again plunge into the sea with a helpless and tremendous fall, which causes the surrounding waters to shoot up in broad and lofty volumes capped with foam.

"Others of the same 'school' leap or 'breach' in a less degree, sportively brandish their broad and fan-shaped flukes in the air, or protrude their heads perpendicularly above the waves like columns of black rock."

Captain Scott, R.N., once told me, as an example of the height to which a cachalot will leap in these gambols, that when standing on the deck of a man-of-war, he has seen the horizon under the animal before it fell again into the water. When we recollect that these whales often reach eighty feet in length, we may appreciate the force which is exerted in projecting this huge mass to such a height.

Here are two instances of humor as exhibited by birds, and showing playfulness without any desire to cause personal annoyance:

Two ladies were sitting at work in a room in which was a pet canary belonging to one of them. The bird threw a reel of cotton on the floor, took the end of the thread in its beak, and wound it first round the neck of one lady and then round that of the other, until the reel was empty, when the bird perched on a chair, and seemed quite pleased with its freak.

The lady to whom the bird belonged tried to unwind the thread from her neck; but every time that she attempted to remove it, the canary flew at her and flapped its wings in her face, so as to prevent her from freeing herself.

A young lady, who was considered as the mistress of a bull-finch by every one except the bird himself, sends me the following account of her pet's sense of humor, which was developed, as is usual among the lower animals, in the form of a practical joke:

"One day while 'Bully' was flying about the dining-room mamma went out of the room, leaving on the table her work-box, in which was a little pincushion stuck full of needles already threaded, besides pins. Bully knew that he was not allowed to touch the pins; but as soon as mamma left the room he pulled all the pins out of the pincushion, and scattered them about the tablecloth. The needles he carried to the top of the lamp, and twisted the thread round a part of the lamp. He then put the pincushion under the sofa, and retired to the lamp, where he waited until mamma came back."

It is really a remarkable fact that these two birds, each belonging to the finches, should have played similar practical jokes with thread, i. e., twisting it about some object where it had no business to be. I ought to add that the story of the canary was sent to me only a few days ago, while no less than thirteen years have elapsed since the anecdote of the bull-finch was sent. It had been put away in a box, and was only found an hour or two before this present time—11.25 A.M., January 8, 1874.

In Bennett's "Wanderings in New South Wales" there is a singularly interesting history of the life and death of a Siamese ape, which went by the name of "Ungka." He was a playful animal, and sometimes, when he could not find a human playfellow, he would try to make companions of some small monkeys that were on board the same ship. He was too big and black for them, however, and they united together for the purpose of driving him away.

"Ungka, thus repelled in his kind endeavors to establish something like sociality among them, determined in his own mind to annoy and punish them for their impudence. So the next time that they united as before in a body on his approach, he watched the opportunity, and, when one was off his guard, seized a rope, and, swinging toward him, caught him by the tail and hauled away upon it, much to the annoyance of the owner, who had no idea that such a retaliation was to take place. He continued pulling on it as if determined to detach it, until the agility and desperation of the monkey at being so treated obliged him to relinquish his hold.

"But it not unfrequently happened that he made his way up the rigging, dragging the monkey after him, and thus made him follow his course most unwillingly. If, in his ascent, he required both hands, he would pass the tail of his captive into the prehensile palm of his feet. It was the most grotesque scene imaginable, and will long remain in the remembrance of those who witnessed it. It was performed by Ungka with the most perfect gravity of countenance, while the poor suffering monkey grinned, chattered, and twisted about, making the most strenuous efforts to escape from his opponent's grasp.

"His countenance, at all times a figure of fun, now had terror added to it, increasing the delineations of beauty; and when the poor beast had been dragged some distance up the rigging, Ungka, tired of his labor, would suddenly let go his hold of the tail, when it would require some skill on the part of the monkey to seize a rope in order to prevent his receiving a compound fracture by a rapid descent on deck. Ungka, having himself no caudal extremity, knew well that he was perfectly free from any retaliation on the part of his opponent."

After this the small monkeys had a consultation, and agreed that whenever Ungka assaulted either of them, they should all unite in attacking him. This, by the way, is another proof of language and power of combination among animals, as mentioned in Chapter V. Having acquired a taste for tail-pulling, and finding it unsafe to attack the monkeys, Ungka took to pulling the tail of the ship's pig, trying, but in vain, to straighten it.

We shall hear something more of this animal in connection with the subjects of Pride and Conscience.

Some persons have asked me what use the capacity of humor could subserve in the next world? I suppose, much the same that it subserves in this. For my own part, I should extremely regret, were it possible, to lose my sense of humor whenever it may please my Maker to summon me into the spirit-world. There are some, even in this world, in whom the sense of humor seems to be absolutely wanting, and, however estimable they may be in character, they are just solemn prigs. I should be sorry to resemble them in the next world.

CHAPTER XI.

PRIDE, JEALOUSY, ANGER, REVENGE, TYRANNY.

PRIDE, or Self-esteem, among Animals.—Etiquette in the Cow-shed.—Pride of Ancestry in the Mule.—Different Positions of the Horse and the Ass among Mules.—The "Bell Mare" and her Value to Muleteers.—Animals Sensitive to Ridicule.—" Pret's" Objection to Disparaging Remarks.—Dislike of Ridicule Shown by "Ungka."—Pride in Personal Appearance.—The Peacock, the Turkey, the Whidah-bird, and the Bird-of-Paradise.—JEALOUSY and its Developments.—Definition of the Two Kinds of Jealousy.—Jealousies between Pets.—"Zeno," "Diver," and their Aquatic Rivalries.—Jealousy of a Dog, accompanied by ANGER, and followed by REVENGE.—The Brown Mouse and its White Rival.—Jealousy and Delayed Revenge among Poultry.—The Love-drama of the Mandarin Duck, and its Adaptability to the Stage.—Comparison with Human Beings.—TYRANNY among the Animals.—Tyranny in the Tiger-beetle.—The Hermit Crabs and their Conflicts.—Tyranny among Gregarious Animals and Birds.—The Tyrant and the Harem.—Comparison with Polygamous Man.—Tyranny in the Aviary.—Tyranny of Pets.—"Duchie" and her Mistress.

THE five characteristics which head this chapter are not pleasing; but, such as they are, they belong to the immaterial, and not the material, part of man. We shall see that the lower animals also possess these qualities, and the inference to be drawn from that fact is obvious. We will take those characteristics in order.

Pride, or self-esteem, is developed as fully in many animals as it can be in the proudest of the human race. This is shown most conspicuously in animals which herd together. There is always one leader at the head, who will not suffer any move to be made without his permission, and who resents the slightest interference with his authority. Especially is this the case with the deer tribe, the horses, and the oxen. Even when these latter animals are domesticated, and the habits of their wild life are materially changed, the feeling of pride exists to the fullest extent.

I have often amused myself by watching the inhabitants of a farm-yard, and seeing how the cows have their laws of precedence and etiquette as clearly defined as those of any European court. Every cow knows her own place and keeps it; she will not condescend to take a lower, and would not be allowed to take a higher. When a newly bought calf is first introduced to the farm-yard, it is treated just like a new boy at school. The previous inhabitants of the yard come and inspect it contemptuously, they decline its society, they crowd it away from the hay-racks; and a new-comer in a farm-yard has about as much chance of approaching the rack at feeding-time as a new boy has of getting near the fire on a cold winter day.

However, as time goes on, the young calf increases in growth, and is allowed to mix with her companions on tolerably equal terms. Then, if a younger animal than herself be admitted, it is amusing to see with what gratification she bullies the new-comer, and how much higher she seems to rank in her own estimation when she is no longer the junior. Should the fates be propitious, she arrives at the dignity of being senior cow, and never fails to assert that dignity on every occasion. When the cattle are taken out of the yard to their pasture in the morning, and when they return to it in the evening, she will not allow any except herself to take the lead. I have heard of one instance where the man in charge of the cows would not allow the "ganger," as the head cow is often called, to go out first. The result was that she refused to go out at all; and, in order to get her out of the yard, the man had to drive all the other cows back again, so that she might take her proper place at their head.

In great portions of this country we make little use of the mule, and its real disposition is not generally known. Those, however, who have been forced into long companionship with this animal have always observed some very curious traits of character in it. Judging from popular ideas respecting the mule, we might think that the animal had no pride in its composition; whereas it is in reality a very proud animal, and fond of good society, as is shown by the following extract from Froebel's "Seven Years in Central America:"

"From drivers and muleteers we may pass to

mules, which are in many respects far more interesting than the former, and whose natural disposition is an attractive subject to the observer of nature.

"One of the most striking characteristics of the mule is his aversion to the ass, and the pride which he takes in his relationship to the horse, which instincts are met with obtrusiveness in the ass and by indifference in the horse. If an ass at any time, urged by the vanity peculiar to its race as related to the mule, happens to fall in with a drove of mules, he will, in all probability, be kicked and lamed by his proud relations. A horse, on the contrary, takes a distinguished position in a drove of mules. The latter crowd around him and follow his movements, exhibiting a violent jealousy, each striving to stand nearest to their distinguished relative.

"This instinct is employed to keep together a drove of mules on a journey or at pasture, by putting a mare to the drove, with a bell around her neck, and called the 'Bell Mare;' by the Mexicans, 'Layegua Madre,' *i. e.*, the mother-mare. This animal is led day and night by a cord, and the whole drove is thus kept under control, and will not leave their queen. It is, therefore, very difficult to separate the drove. The man who leads the mare is instructed, in case of an attack from the Indians, to leap instantly upon the back of this animal, and take refuge in the wagon encampment, whither the drove is sure to follow him.

"Even if the Indians succeed in separating any mules from the drove, they find it difficult to carry them off. The animals incessantly attempt to turn back, and the travelers are thus enabled to overtake the robbers, and recover the stolen animals. The Indians, in consequence, use every means to get possession of the mare, and, if they succeed in this, the whole drove is lost to their owners. If several horses are in a drove of mules, the danger is that the latter become dispersed; and this is the reason that, in these journeys, saddle horses are not allowed to go loose, but are led by a cord."

It is rather curious to trace among the lower animals a feeling which bears a very close resemblance to pride of birth among mankind. Pride shows itself in many ways, both in men and animals. Here we have pride of rank and love of precedence among cows, and pride of ancestry among mules. Sometimes pride takes the form of sensitiveness to ridicule. There is nothing so galling to a proud man as to feel himself the object of ridicule, and precisely the same trait of character is to be found in many animals.

As may be expected, this form of pride is mostly developed in the domesticated animals; or, perhaps, it is in those that we have most opportunities of observing it.

My cat, "Pret," for example, was peculiarly sensitive to any thing approaching ridicule. He was quite conscious if we spoke of him in a disparaging manner, and testified his disapprobation after his own manner. But to laugh at him was an insult which he could not brook, and, if we continued to do so, he would arch his tail, hold himself very stiff indeed, and march slowly out of the room. How sensitive all high-bred dogs are to ridicule is so well known that we need not occupy space by citing examples.

The Siamese ape "Ungka," a part of whose history has already been given in connection with the subject of Humor, possessed a keen sense of ridicule. The animal was exceedingly tame, and at meal-times always came to take his share, a corner of the table being appropriated to his use. "When, from any of his ludicrous actions at table, we all burst out in loud laughter, he would vent his indignation at being made the object of ridicule by uttering his peculiar hollow barking noise, at the same time inflating the air-sac, and regarding the persons laughing with a most serious look until they had ceased, when he would quietly resume his dinner."

Pride in personal appearance, or vanity, is often to be seen among the lower animals, more especially among those birds who are notable for bright or abundant plumage.

Any one who has seen a peacock in all the glory of his starry train will recognize the intense pride which the bird feels at his own splendor. He does not only display his magnificent train for the purpose of attracting the homage of his plainly clad mates, but seems to be just as proud of the admiration bestowed by human beings as of that offered by his own kind.

Nor does he despise the homage of birds whom he considers his inferiors. Only a few hours before writing these lines I saw a peacock, with his train fully spread, displaying all his grandeur around a dozen barn-door fowls. He stalked majestically among them, scarcely deigning to look at them, but turning round and round so as to display his grandeur to the best advantage, and apparently as satisfied with the effect which he produced as if he had been surrounded by his own kind.

Then there is the turkey, whose movements are so grotesque when he is strutting about in his nuptial plumage, and who surveys himself with ludicrous complacency. Taking the well-known whidah-bird (or widow-bird, as it is often called), we see this trait of character highly developed. He is wonderfully proud of his beautiful tail, and, as long as he wears it, loses no opportunity of displaying it to every visitor who approaches his cage. But when the moulting season comes, and he assumes for a while the plain, tailless suit of his mate, his manner is as changed as his appearance, and, instead of exhibiting himself in all his pride, he mopes about with a dull and listless demeanor, and seems absolutely ashamed of his mean condition.

It might be expected that so magnificent a creature as the bird-of-paradise would have its full share of pride; and that this is the case is shown by the account of a tame specimen in Bennett's "New South Wales," to which reference has already been made:

"One of the best opportunities of seeing this splendid bird in all its beauty of action, as well as display of plumage, is early in the morning, when he makes his toilet. The beautiful sub-alar plumage is then thrown out, and cleansed from any spot that may sully its purity by being passed gently through the bill. The short chocolate-colored wings are extended to the utmost; he keeps them in a slowly flapping motion, as if in imitation of their use in flight; at the same time raising up the delicate long feathers over the back, which are spread in a chaste and elegant manner, floating like films in the ambient air.

"In this position the bird would remain for a short time, seeming proud of its heavenly beauty, and in raptures of delight with its most enchanting self. It will then assume various attitudes, so as to regard its plumage in every direction.

"I never yet beheld a soil on its feathers. After expanding the wings, it would bring them together so as to conceal the head: then, bending gracefully, it would inspect the state of its plumage underneath. This action is repeated in quick succession, uttering at the same time its croaking notes. It then pecks and cleans its plumage in every part within reach; and, throwing out the elegant and delicate tufts of feathers underneath the wings, seemingly with much care and not a little pride, they were cleaned in succession by throwing them abroad, elevating, and passing them in succession through the bill.

"Then, turning its back upon the spectator, the bird repeats the actions above mentioned, but not in so careful a manner; elevating its tail and long shaft feathers, it raises the delicate plumage of a similar character to the sub-alar, forming a beautiful crest, and, throwing its feathers up with much grace, appears as proud as a lady in her full ball-dress. Having completed the toilet, it utters the usual cawing notes, at the same time looking archly at the spectators, as if ready to receive all the admiration that it considers its elegant form and display of plumage demand. It then takes exercise by hopping in a rapid but graceful manner from one end of the upper perch to the other, and descends suddenly upon the second perch, close to the bars of the cage, looking out for the grasshoppers which it is accustomed to receive at this time."

Here we have the character of pride in personal appearance developed as strongly as it could be in any human being. Moreover, the bird could sufficiently enter into the feelings of the spectators to understand that they were admiring its beauty, and to exult in that admiration.

JEALOUSY.

Jealousy is of two kinds, one connected with the love of some other being, and the other depending on the love of self. The former is thus defined in Webster's Dictionary: "That passion or peculiar uneasiness which arises from the fear that a rival may rob us of the affection of one whom we love, or the suspicion that he has already done it." The latter is thus defined: "The uneasiness which arises from the fear that another does or will enjoy some advantage which we desire for ourselves." We will only deal with the former of these traits of character.

In the first place, it is evident, from the definition which has just been given, that jealousy implies the power of reasoning, and that any creature by which it is shown must be able to draw a conclusion from premises. Perhaps the animal is wrong in its conclusion; but the process is still one of reasoning, however incorrect that process may be.

All persons who have possessed pet animals must be familiar with the exceeding jealousy displayed by most of them. This feeling is manifested most strongly when an animal has been the only pet, and another is introduced into the house.

My own cat, "Pret," resented so strongly the advent of a Skye terrier, that when the dog came into the house he walked out, and never would enter it again. He had already put up with a baby, which was a very great trial to his feelings; but a dog was more than he could endure, and so he retired to his own house in the garden, and lived there alone. His affection for me remained unchanged, and he was only too happy when I went into the garden. But he held the house desecrated by a dog, and, even when hungry, could not be allured within the door by the offer of food.

Not that he had any personal objection to the dog; on the contrary, the two animals were very friendly with each other, even eating out of the same dish. But Pret considered that "Bosco" had no right to me, and whenever I came on the scene, Bosco got his ears boxed, and had to retire into the background. So absurdly strong was this jealous feeling, that whenever I wanted Pret to come quickly, I used to call Bosco; which ruse always had the effect of bringing Pret along at full speed, in order to anticipate the dog.

One of my neighbors has a couple of little dogs— "Bell," a black-and-tan toy terrier, and "Fay," a cross between Skye and Maltese. These two animals are the best of friends, always lying on the same mat, which they share with an enormous cat called "Tommy." But, with regard to their human friends, their jealousy of each other is extreme. They do not seem to care if notice be taken of the cat; but if Bell be caressed, Fay is sure to sidle up and try to interpose herself between Bell and the caressing hand. Bell is equally jealous, but shows her feeling by noisy and angry demonstrations of assault, which, however, are never carried into effect.

I suppose that most of my readers who have possessed two or more dogs at the same time must have been amused at the boundless jealousy which they will display toward each other while engaged in the service of their master, though at other times very excellent friends. Such scenes as the following are of frequent occurrence, and are instructive as well as amusing:

"I have in my possession a favorite dog called 'Zeno' (from the Greek philosopher); his age is over seventeen years; he is a cross between an otter terrier and a Scotch.

"There never was a cleverer dog. He is passionately fond of the gun; and though very serviceable in the field, still I disliked to have him when pointers were at work. He was the finest retriever I ever saw, and if there was a wounded bird, hare, or rabbit, if he were allowed, he would find it, go where it might be, even into the sea. Wherever I went to shoot, he was sure to find me, no matter whether I walked, drove, or rode. I have seen him come up to me of an afternoon, when he must have traveled seven or eight miles, and well did he know he was doing wrong. He would sneak up behind me, afraid almost to show himself; but whenever I gave him a sign of kindly recognition he became quite frantic and overjoyed; in an instant, however, he was at my heels.

"A favorite amusement of mine was to shoot wild-fowl and teal in the Frith of Forth. I used to take Zeno and a large Newfoundland, named 'Diver,' with me.

"Zeno was ever on the watch, and, wherever I killed the bird, it was amusing to see the two plunging into the waves, and racing to get first to it. Zeno generally picked up the bird, having no heavy coat to impede him; but Diver often thought that he should have the honor of carrying it, and would attempt to take it from him— but it was of no use. The battle often became fierce, the little dog dropping his game, flying at the larger one with all his fury, then, picking up his bird, would paddle his way to the boat-side, look up in my face as if saying, 'Have I not done well?' and then I would help him up, when the two were as good friends as ever."

The power of jealousy and anger is well shown by the following anecdote of a pet dog. I knew the animal well, and his mistress wrote the little history at my request:

"One of our pet dogs, of a mixed breed, with long white hair, was, in common with most pets, of a very jealous disposition, always showing displeasure if any other living thing obtained a share of that attention which he considered exclusively his own.

"One wintry morning a poor little infantine duck that had been injured was brought into the house to be nursed and tended. The dog watched all the attentions bestowed upon it, was evidently annoyed at the intrusion of a rival where he had ruled supreme, and vowed vengeance.

"After a few days the duck mysteriously disappeared. The dog was suspected, but neither dog nor duck could be found. Just before dark a more minute search was made in the house and garden, and at last something was visible under a large rose-tree. There stood the culprit, shivering with cold, his nose and paws all covered with mud, and at his feet was a half-filled grave, in which was deposited the body of the murdered duck.

"His long hair had become entangled in the thorns of the rose-tree while he was engaged in burying his victim, and fear of detection and reproof had caused him to remain a silent captive for so many hours. His pitiful condition disarmed our censure, for he was so firmly fixed that the gardener had to cut off the rose-branch, so that it might be more leisurely disentangled in the house. Before this event the gardener had not been a favorite with the dog, but ever afterward it seemed to feel itself owing a debt of gratitude to its deliverer."

The dog in question lives at Canterbury, where his intellect and accomplishments have made him well known.

We see in the behavior of the animal, not only jealousy, but memory, hatred, and revenge, and a sense of moral responsibility. The remembrance of the favors lavished on his rival rankled in his memory, and the result was hatred culminating in revenge when he found an opportunity. Then he knew that he had done wrong in killing the duck, and, just as a man would do who had committed a murder, tried to conceal the evidences of his crime by burying the body of his victim. So deeply was his conscience pricked, that, when he found himself arrested by the bush, he ran the risk of dying of cold and hunger rather than allow himself to be discovered.

Even in such rarely tamed animals as the common mouse the feeling of jealousy has been known to be so strong as to lead to murder. A young lady, one of my correspondents, had succeeded in taming a common brown mouse so completely that it would eat out of her hand, and allow itself to be taken off the floor. She had also a tame white mouse in a cage.

One morning, when she went to feed the white mouse as usual, she found it lying dead on the bottom of the cage, and beside it was its murderer, the brown mouse. The cage being opened, the latter made its escape, but how it had contrived to gain admission was a mystery.

An instance has lately come to my knowledge where jealousy was restrained for a considerable time through fear, and at last broke out when the cause of fear was removed. The story is told by a lady living in Edinburgh:

"I remember a Malay cock of mine manifesting a mixture of hatred and revenge to a dead rival, equal in fury, if not in power, to what a Malay man, in similar circumstances, might have shown.

"We had a very splendid dunghill cock, who kept the Malay (a cowardly caitiff) in great subjection. This cock died suddenly. His rival came by chance on his dead body. He instantly sprung on it, kicked, spurred, and trampled upon the lifeless bird, and, standing upon the corpse, flapped his wings, and crowed himself hoarse with the most disgusting energy.

"The rascal took instant possession of the harem, and I often thought that the hens must sadly have missed their old lord. He always used to share any titbits with them—a practice carefully avoided by his successor, the Eastern despot, who greedily kept the best to himself."

Again, comparing man with beast, we see that the bird in this case acted exactly as a savage does when his enemy has fallen. The savage exults over the dead body of an enemy, especially if that foe has been very formidable in life, and mutilates in futile revenge the form which he feared when living.

Take the following story, which is related in Bennett's "Wanderings in New South Wales," transform the actors into human beings, and see how exactly the birds acted like human beings, and how the plot of a powerful drama might be constructed from the story. The birds in question were the beautiful little "Mandarin" ducks, which even in China are exceedingly valuable. They are proverbial for their conjugal fidelity, and in marriage processions a pair of these ducks are carried about as emblems of the love which ought to animate the newly married couple.

"The following circumstance of fidelity was mentioned to me as having occurred in two birds of this species:

"A drake was stolen one night, with some other birds, from Mr. Beale's aviary. The beautiful male was alone taken, and the poor duck was left behind. The morning following the loss of her husband, the female was seen in a most disconsolate condition; brooding in secret sorrow, she remained in a retired part of the aviary, pondering over the severe loss she had just sustained.

"While she was thus delivering her soul to grief, a gay, prim drake, who had not long before lost his dear duck, which had been accidentally killed, trimmed his beautiful feathers, appeared quite handsome, and, pitying the forlorn condition of the bereaved, waddled toward her, and, after devoting much of his time and all his attention to the unfortunate female, he offered her his protection. She, however, refused all his offers, having made, in audible quacks, a solemn vow to live and die a widow if her mate did not return.

"From the day on which she met with her loss she neglected her usual avocations, forsook her food and usual scenes of delight, where she loved to roam with him now absent, and to excite his brave spirit to drive away all the rivals that might attempt even to approach them. But those fleeting hours of enjoyment had passed, perhaps never to return, and no consolation that could be offered by any of her tribe had the least effect. Every endeavor was made to recover the lost bird, as it was not expected that the beautiful creature would be killed.

"Some time elapsed after the loss, when a person, accidentally passing a hut, overheard some Chinese of the lower class conversing together. He understood sufficient of their language to find out that they said, 'It would be a pity to kill so handsome a bird.' 'How, then,' said another, 'can we dispose of it?' The hut was noted, as it was immediately suspected that the lost Mandarin drake was the subject of the conversation. A servant was sent, and, after some trouble, recovered the long-lost drake by paying four dollars for him. He was then brought back to the aviary in one of the usual cane cages.

"As soon as the bird recognized the aviary, he expressed his joy by quacking vehemently and flapping his wings. The interval of three weeks had elapsed since he was taken away by force; but when the forlorn duck heard the note of her lost husband, she quacked, even to screaming with ecstasy, and flew as far as she could in the aviary to greet him on his restoration. Being let out of the cage, the drake immediately entered the aviary, and the unfortunate couple were again united. They quacked, crossed necks, bathed together, and then are supposed to have related all their mutual hopes and fears during the long separation.

"One word more on the unfortunate widower who kindly offered consolation to the duck when overwhelmed with grief. She, in a most ungrateful manner, informed her drake of the impudent and gallant proposals he made to her during his absence. It is merely supposition that he did so; but, at all events, the result was that, on the day following his return, the recovered drake attacked the other, pecked his eyes out, and inflicted on him so many other injuries as to occasion his death in a few days. Thus did this unfortunate drake meet with a premature and violent death for his kindness and attention to a disconsolate lady. It may perhaps be correctly written on his grave, 'A victim to conjugal fidelity.'"

The very same feelings which would have actuated human beings under similar circumstances influenced these birds. There is conjugal love affected by sudden and violent separation; there is conjugal fidelity in absence; there is sorrow for the loss of one who is loved; there is joy in reunion; there is jealousy at an attempt to steal the affection of a partner; and, finally, there is revenge swiftly taken upon the offender. There also is the power of language, as, without a very definite language of her own, the duck could not have told her partner of a particular drake, and so drawn upon him the vengeance of her husband.

TYRANNY.

Another of the many traits of character which are common to man and the lower animals is tyranny, *i. e.*, the oppression of the weak by the strong, whether that strength belong to the body or the mind. In many of the animals, tyranny takes its most obvious form, the strong not only oppressing the weak, but killing and eating them, even though they be of the same species. A human cannibal acts just in the same way, eating his enemy after he has killed him.

As to the milder forms of tyranny, there is scarcely an animal in which it may not be found, and it is manifested quite strongly in the insects. There is a notice in Hardwicke's *Science Gossip*, for October, 1871, of tyranny shown by a tiger-beetle toward its fellows, one insect assaulting another, and driving it away, "much in the same manner as sparrows do when they have secured some morsel of food which they wish to keep to themselves."

In the *Daily News* of November, 1873, there was a brief and amusing account of tyranny as shown by crabs. The writer had been observing the proceedings of the creatures in the Brighton Aquarium :

"It is well worth while to watch the movements and manœuvres of the hermit crab. He is discerning, has a keen eye to his own convenience, pugnacious when any one comes between him and the object of his desire, and unrelenting in following up his advantage. He contends for some practical and substantial end, pursuing conquest not for the sake of the bare submission of his adversary.

"These remarks are induced by our own observation of the amusing habits of this little animal. Some time back we were, one bright morning, watching the beautiful gleam of the herring, as its scales reflected the sunshine in varied colors, which played into one another, reminding one of a beautiful 'shot' silk dress. Our attention was suddenly attracted by a commotion among the hermit crabs, many of which are in the same tank with the herrings.

"These crabs, as is generally known, have recourse to the cover of a whelk, or other shell, to protect the abdomen, which is very soft and vulnerable. Suddenly one of the number, a large specimen, whose adopted dwelling was of somewhat narrow dimensions, gave chase to a small crab which occupied a shell much larger than that of his bigger neighbor. The little one, apparently quite alive to the sinister intentions of his pursuer, took to flight as quickly as possible, and his attempts to escape were continued with the utmost vigor until further effort was hopeless. The way in which he dodged around and behind oysters, and whatever afforded him a temporary cover, was amusing in the extreme.

"At length he was overtaken, and then a regular pitched battle ensued. The little one resisted manfully, but was finally overcome, the more bulky combatant having, after the most strenuous exertions, succeeded in forcing his claws between the body of his weaker opponent and his adopted shell, and with the most frantic exertion turning him out. They then, apparently as a matter of course, exchanged shells, the ousted tenant yielding submissively to his fate, and quietly adapting himself to his reduced circumstances. In this encounter, from beginning to end, all the qualities we began by enumerating were exemplified in these little creatures—the discernment with which the larger crab fixed upon the shell which exactly suited him, the determination with which he followed up his intention of possessing himself of it, and the pugnacity and perseverance displayed by both in the course of the encounter.

"This was evidently no fight for mere fighting's sake; but the whole proceeding evinced a settled plan, pursued, on the one hand, with the greatest determination, and, on the other hand, met with the most obstinate resistance."

I have often witnessed similar scenes, not only in the Aquarium, but in the rock pools along the coast.

Tyranny is invariably seen among gregarious animals, the herd or flock being always under the command of one individual, who has fought his way to the front, and who will rule with imperious sway until he becomes old, and in turn is ousted by a younger and more vigorous rival. The same quality is very familiar to us in our poultry-yards, where, no matter how many may be the number of birds, one cock invariably assumes the leadership.

As a rule, he takes his honors meekly, but bases his conduct on the old Roman principle, "*Parcere subjectis, sed debellare superbis.*"

There are mostly one or two younger cocks, with whom he does not interfere, unless they attempt to dispute his sway, or—most unpardonable offense of all—to win the affections of any of his harem. In such cases an immediate attack is the result. If he win (as he generally does, if only by reason of his *prestige*), the state of society remains unaltered. But, if he lose the battle (which generally means losing his life), the conqueror succeeds to his place, and takes as a matter of right all his possessions, including his harem.

It is curious to trace the analogy between these birds and human beings, especially those of the East, whether at the present day or in the ancient times, as depicted in the Old Testament. Substitute human beings for birds, and the country at large for the poultry-yard, and the resemblance is exact. There are many petty chieftains; but among them is sure to be one more mighty than the rest, who holds his place by superior force, either of intellect or military power. If challenged by one of the inferior chiefs and is victorious, he retains his post; while, if he is vanquished, his conqueror takes his place, his property, and his wives. And it is another curious point that, whether with men or birds, the members of the harem seem to trouble themselves very little about the change of master.

The Scriptures are full of allusions to the invariable custom that the conqueror takes the possession of the harem belonging to the vanquished. David did so with regard to the women of Saul's household. "I anointed thee king over Israel, and I delivered thee out of the hand of Saul. And I gave thee thy master's house, and thy master's wives into thy bosom" (2 Sam. xii. 7, 8). So, when Nabal died after his defiance of David, the latter, as a matter of course, took possession of Nabal's wife, together with the rest of his property. Similarly, as had been foretold by the prophet Nathan, when Absalom rebelled against David, he publicly took possession of his father's harem as a sign that he had assumed the kingdom.

To those unacquainted with Oriental customs, it seems strange and cruel that when Adonijah asked Bathsheba to persuade her son Solomon to give him Abishag as a wife, he should be at once put to death. But, as explained by those customs, he had for the second time committed high-treason. He was Solomon's elder brother, and had already made an attempt to gain the throne. He had failed, and had been pardoned on condition of future good conduct. But his demand for the hand of Abishag, who had belonged to David's harem, was considered equivalent to a second rebellion; and so he and his principal supporters, Joab and Abiathar, met with instant punishment, the former with loss of life, and the latter with deprivation and banishment.

The whole scene is worthy of notice. Bathsheba visits her son in full court, and prefers her request. Solomon, who treated her with the greatest respect as the king's mother, instantly treats Adonijah's demand as an overt act of high-treason. "And King Solomon answered and said unto his mother, And why dost thou ask Abishag the Shunammite for Adonijah? Ask for him the kingdom also, for he is mine elder brother; even for him and for Abiathar the priest, and for Joab the son of Zeruiah."

Now Solomon was a man remarkable for his mercy in an unmerciful time and among a ruthless race. He was probably the only Oriental monarch who would not have secured himself on the throne by putting to death all his brethren— a custom which prevails to the present day. Yet he not only spared Adonijah's life, but forgave him after actual rebellion. This second offense was, however, unpardonable, the demand of David's wife being tantamount to a claim on David's throne, and therefore he paid the penalty with his life, as being a dangerous man who could not be trusted. Besides these instances, there are many other allusions to the custom scattered through the Scriptures.

The closeness of the parallel between man and beast is most remarkable. In both there is a single despotic ruler who allows no rival; and in both an attempt to gain the affections of one of the harem is considered tantamount to a challenge for sovereignty, and is treated accordingly.

Sometimes a very curious sort of tyranny is shown where a number of creatures are confined in the same place. Mr. Bennett has some remarks on this subject in connection with the Mandarin ducks which have already been mentioned under the head of Jealousy.

Speaking of the feeding-time in the aviary, he says: "It is at this time that we can also observe the querulous disposition of these animals. The males of one and the same kind of a different species endeavor to grasp all the supplies for themselves, unmindful of the wants of others, and will not even permit their companions to perform their ablutions without molestation, although they may have themselves completed what they required.

"I often observed the Mandarin ducks excite the drakes to attack other males or females of the same species, as well as any other kind of bird (not too powerful) in the aviary, against whom the lady may have taken a dislike from some cause or causes unknown to us. There always appear to be one pair who exercise a tyranny over the others, not permitting them to wash, eat, or drink, unless at the pleasure of these little aristocrats."

But, of all tyrants, commend me to a spoiled dog, who is even worse than a spoiled child. Obedience is a thing unknown to him. If he is wanted to go out for a walk and prefers to stay at home, he stays at home, and his master goes out alone. If he wants to go for a walk, he makes his master go with him, and take the directions which he prefers. Perhaps a better example of tyranny has never been given than Dr. J. Brown's history of the Skye terrier "Duchie." The little animal so completely domineered over her mistress that the latter could not even choose her own dinner, but was obliged to have whatever Duchie preferred, and was once kept out of bed for half a winter's night because Duchie had got into the middle and declined to move.

CHAPTER XII.

CONSCIENCE.

Definition of Conscience.—Its Exercise by the Lower Animals, and Comparison with the Conscience of Man.—Sense of Moral Responsibility in the Dog.—The Butcher's Dog, his Master, and the Old Woman's Money.—Parallel between Dog-servants and Human Servants.—Voluntary Responsibility.—"Vic" and the Croquet-hoops.—Zeal outrunning Discretion.—The Robber in Custody of "Help."—Dog Cooks and Nursemaids.—"Offy" Saving the Servant's Life.—Duty Paramount.—"Bree's" Fearful Leap.—Mistaken Ideas of Duty.—Church-going Dogs.—"Apollo's" Leap.—"Boxer's" Notion of Duty.—Epigram from "Salmagundi."—Sin and Penitence.—Suicide or Sorrow.—A Dog's Grave.—A Dog's Sense of Wrong-doing.—Guardianship.—The Cat and the Butcher.—"Ungka's" Theft and Restitution.—"Tokla," the Hunting-dog, and the Sheep.

To those who have never studied the ways of the lower animals, it may seem strange to assert that they, as well as ourselves, possess conscience, *i. e.*, a sense of moral responsibility, and a capability of distinguishing between right and wrong. It is necessarily developed strongest in those animals which are placed under the rule of man, and especially in those which, like the dog, belong to his household, and are made his companions. Conscience, in their dealings with man, is their religion, and they often exercise it in a way which would put many a human being to shame.

It is this feeling which induces the dog to make itself the guardian of its master's property, and often to defend that property at the risk of its life. For example, if a dog be placed in charge of its master's dinner, the faithful animal will never touch a morsel of food, however hungry it may be. Nay, a dog would rather, as an ordinary rule, die of starvation than eat the food which belonged to his master. We often see field-laborers working at one end of a large field, while their spare clothes and their dinner are at the other end, guarded by a dog. They are quite easy about the safety of their property, knowing well that the dog will not allow any one to touch either the clothes or the provisions.

A still stronger instance of moral responsibility in a dog has just come before me. A poor woman, who lived in an unprotected part of Scotland, became unexpectedly possessed of a large sum of money, with which property she was as much troubled as "Captain Jack" with the money which he dared not spend, was afraid to show, and could not carry about him for lack of pockets. She would have taken it to the bank, but could not leave the house.

At last she asked the advice of a butcher of her acquaintance, telling him that she was afraid to live alone in the house with such a sum of money.

"Never fear," said the butcher; "I will leave my dog with you, and I'll warrant you that no one will dare to enter your house." So toward evening the dog was brought, and chained up close to the place where the money was kept.

In the middle of the night a robber made his way into the house, and was proceeding to carry off the money, when he was seized by the dog, who held him a prisoner until assistance came. The thief was the butcher himself, who thought that he had made sure of the money. He had not considered that his dog was a better moralist than himself, and, instead of betraying a defenseless woman, would even take her part against his own master.

The woman kindly pardoned the intending robber; and I hope that for the future he took a lesson from his own dog, and amended the evil of his ways.

A rather notable instance is now before me, where the capacity of conscience, as it is manifested in the lower animals, is very well shown. There is a retriever belonging to a Scotch lawyer, who was a very conscientious animal in his way; that is to say, as far as his intellect would carry him, he was absolutely conscientious; but, unfortunately, there was a limit beyond which conscience could not assert itself.

For example, no matter how hungry he might be, the dog might be safely left in a room where the dainties which he best loved were left unguarded: not a morsel of food would he touch. But he did not offer any objection to the cat when *she* stole the food from the table; neither did he display any scruples in sharing with her the product of her theft.

G

Neither was he conscientious enough to submit passively to imprisonment when his master wished to dispense with his company : he had a knack of gnawing cords asunder, forcing shutters and opening doors, which showed to a certainty that the animal was actuated in such matters by reason, and not by instinct.

This faithful animal was killed suddenly by a railway train. His master writes of him as follows : " He was the most honest animal I ever knew, and I only wish that we could get servants as honest. Upon the whole, I think that he was a much more exemplary character than many men and women whom I have known, and I should be very happy to meet him again in some other sphere. I would rather hunt with him on a planetoid, or a ring of Saturn, than spend my time in the narrow heaven which some zealots would arrogate to themselves and their small sect, if they could. He certainly had much more charity than they."

Not only does the dog guard the property which is intrusted to its charge, but it often goes a little further and assumes a charge on its own account.

I was lately staying with a friend in the country, and became on very excellent terms with his little bull-terrier, named " Vic." On the second day after my arrival a croquet match was arranged, and I was asked to change the position of the hoops, so as to suit the latest development of the game. Accordingly, I went to the lawn, followed by Vic, who took no particular notice, but lounged about the lawn with no apparent object.

Presently my friend joined me. " Do you know," said he, "why Vic is loafing about here?"

" No, except that she prefers the garden to the house."

" Not a bit of it. She has come to see that you do not take away any thing out of the garden; and so I came to warn you not to take a hoop or a peg off the lawn."

It appeared that she always acted in the same manner toward people whom she did not know intimately, although, after a time, she had confidence and let them alone. In point of fact, after two or three days had elapsed, Vic never troubled herself about me.

On one occasion her fidelity took an unpleasant form. Her master had lent his chaise to a friend, who was driving it, and who came to a hill. He dismounted, and was stooping to put on the skid, when Vic flew at him, having an idea in her head that he was going off with the wheel. One of the oddest points in Vic's conduct is that, as if out of a sense of politeness, she does not make her watch an ostentatious one, but merely keeps in view the object which she is guarding, and the person of whom she is suspicious.

In the cases which I have mentioned, Vic was entirely wrong in her surmises. A remarkable instance, however, has reached me, where the dog was right, and acted in a way that would have been creditable to any human being :

" Of ' Help,' a Newfoundland dog, several stories are told, and there was one especially which showed that his sense of responsibility overcame that of friendship. His master owned a wood-yard, from which there had been a constant series of mysterious thefts. At last the dog was put into the yard for the night, in hopes that he might scare away the thief.

" Next morning, Help was found guarding one of the men belonging to the premises, who had a bundle of wood upon him. The man was aware that the dog knew him perfectly well, and had presumed on the animal's forbearance. Help, however, assumed so fierce an aspect that even the certainty of detection did not give him courage to oppose the faithful creature, nor even to get rid of his compromising load."

Dr. J. Brown relates a similar anecdote of " Rab." He flew at the throat of a man who tried to rob his master, pulled him down, and remained in charge of the fallen man while his master went on his journey. After a while he was seen coming alone to rejoin his master. It appeared that the robber was a neighbor whom the dog knew, and so, giving him a thorough fright, he let him off, after subjecting him to very humiliating treatment.

It is well known that in India the elephant is sometimes taught to take charge of children, especially if they happen to belong to his driver ; but the dog appears to be even a more curious nurse than the elephant. The dog Help, however, who took prisoner the midnight robber, was accustomed to act as nurse, and performed the task as well as any elephant could have done, and indeed better than some nursemaids do.

" At times Help could take the place of the nursemaid, and was often intrusted with the entire charge of a little girl, only old enough to crawl on the floor. As long as she amused herself safely, Help looked on with quiet dignity ; but whenever she moved toward the fire, or in any direction that seemed dangerous, he put his great paw upon her, and turned her another way."

One of my friends has written to say that she knows a dog in Berkshire who acts on the same principle as Help, though, on account of his small size, he can not do without human assistance. When the cook puts a saucepan on the fire, she sets the dog to watch it, and can go about other business in perfect security, knowing that, if the saucepan should boil over, he will call her. He is also put in charge of the cradle; and if the child should wake up, goes and fetches the nursemaid. In these cases, besides the sense of responsibility, there is much reasoning power, and a capability of understanding human language. And it must be noticed that the dog last mentioned never brings the cook to the cradle, nor the nurse to the saucepan.

A somewhat similar case is related by a lady whom I have known for some years:

"One of my earliest recollections is of a fat, black, curly-haired old dog, called 'Offy.' This was an abbreviation of orphan, his mother having died immediately after his birth, and his father being unknown. Offy was the delight of our hearts, the kindest, gentlest, and most attached of dogs. At night he always lay at the top of the stairs leading to the nursery, so that no one could enter without his permission.

"Once, when we were at the seaside, my nurse had a rheumatic fever, and was quite helpless. By chance one day she was left alone, propped up with pillows in a large arm-chair near the fire. Suddenly the house resounded with Offy's barks. One maid was out walking with us, and the other was busy getting dinner in the kitchen, when Offy's barks attracted her notice. Running upstairs, she was met by the dog, who ran down to meet her, caught her by the dress, pulled her up the stairs, and there, lying on the floor, with her head and arm on the fender, lay the poor nurse, unable to move. Without the dog, she must have been burned to death."

Here we have a variety of qualities which certainly do not belong to the mortal part of a living being, whether man or beast. First, there is sympathy with suffering. Then there is reason, telling the animal that the fallen woman was in danger, and could not help herself. Reason also told the dog that he was incapable of helping her himself, and that he must summon some one who had the power. He then had recourse to his own language, which he knew would be understood, and called for help as intelligibly as if he had spoken human words.

Sometimes conscience assumes the form of moral responsibility, the animal being determined, at any risk, to perform the task which is allotted to it. A gentleman, to whom I am indebted for many original anecdotes about various animals, has sent me the following account of determination to fulfill a duty:

"'Bree,' an English water-spaniel belonging to me, was bred from two London prize dogs of that class.

"A few years ago, his former master went to St. Abb's Head for shooting. At the 'Staples Heugh' he winged a duck. The bird, in agony, rolled over the precipice; while poor Bree ran with such impetuosity that he jumped over into the sea, a height exceeding one hundred feet, and fell into some forty feet of water.

"When he came to the surface, there was no place near where he could land, and, seeing the duck rounding the 'West-hare-cars,' he struck out, and, after following it past the 'Skelly,' the 'Ramfands,' the Goose-cruives,' etc., overtook it at the 'Clawfords,' in 'Hare-law-cove Bay.' Grasping the duck in his mouth, he proceeded with it up 'Eel-car-brae,' one of the most difficult passes on St. Abb's Head, and, on reaching his master, laid the bird at his feet."

The distance which the dog swam is somewhere about a mile. The gentleman, through whom this anecdote was sent to me, writes as follows: "As to the story of St. Abb's Head, you would require to see it before you could appreciate what a dreadful leap the dog had taken." I possess photographs of Bree, his master, and St. Abb's Head. The latter word, by the bye, is an abbreviation of Ebba, the remains of whose convent are still to be traced, close to the head itself. "Staples" is a corruption of "steeples," a word derived from some rocks that stand steeplewise in the sea.

The fall must have been a tremendous one, and how the dog escaped instant death is more than I can imagine. The shock must have been a very severe one, and the animal would have been quite justified in coming ashore at once. But he knew that his duty was to catch the duck, and he did so. That he did receive a very great shock is evident from the fact that, although a retriever, and by nature a good water-dog, he has since this adventure contracted such a horror of the sea that he can scarcely be induced to enter it.

Sometimes the dog takes up a wrong idea of duty, but perseveres in it, notwithstanding all obstacles. In the two following instances the dog considered that his duty lay in accompanying his master, and set himself to discover some plan of overcoming obstacles:

"A friend of ours, a clergyman in one of those rural Welsh villages whose name we find some difficulty in writing, and still more in pronouncing, had a spaniel, sent from a friend in England to the rectory of C——. I forget now the correct spelling, but no matter.

"Soon after his arrival, the dog proved himself a most determined church-goer. The first attempt took the family quite by surprise. They knew not he had accompanied them, until they had taken their seats; so they very wisely pushed him underneath, where he remained during the service, one of the quietest members of his master's congregation. The next Sunday, when the church bells commenced, the dog was shut in the library; but, soon after the service had begun, he jumped through the window, pushed open the church door, walked with all proper demureness to his own pew, and resumed his former position under the seat, where he was again allowed to remain. On the third Sunday, the dog's movements were more vigilantly watched. Directly the bells began, he started off full trot to the church, once more occupied his old corner, defied alike the threats and persuasions of the servant to remove him, and, on the arrival of the family, welcomed them triumphantly.

"One more last attempt was made on the succeeding Sunday to keep him away, which was only a partial success. Early in the morning he was shut in a shed, from which he could find no egress; but, directly the bells began, he struck up a loud howling accompaniment, which he continued during the whole of church-time, and, as the church was close to the rectory, he could be heard at intervals during the service, of course disturbing the risible powers of the junior members of the congregation, so that nothing remained but to send him back to his former master in England."

The dog in question ought to have learned by heart an epigram in a curious and very scarce quarto book called "Salmagundi." It was published in 1791, and I have a copy, which was presented by the author to my grandfather. It is, in its way, as interesting as are Gilray's political caricatures, comprising, as it does, the famous "Wilkes and Liberty" times, and abounding with witty little *jeux d'esprit* in Latin and English. Here is the epigram which has been mentioned:

ON A FAVORITE DOG WHO REGULARLY ACCOMPANIED HIS MISTRESS TO CHURCH.

"'Tis held by folks of deep research,
He's a good dog who goes to church;
As good I hold him every whit,
Who stays at home and turns the spit;
For though good dogs to church may go,
Yet going there don't make them so."

A somewhat similar instance occurred to myself. I was making some arrangements in the church, and had left my dogs outside, thinking that they would amuse themselves by swimming in a neighboring pond, as they were accustomed to do. I had, however, not been in the building for many minutes when a scratching, scrambling sound was heard, followed by a heavy thump, and up came my bull-dog "Apollo," looking delighted to see me.

I put him out at the door, but could not imagine how he had made his entrance. Presently there was another scratching, and I saw Apollo's head at a little window which had been left open for ventilation. He contrived, in some curious manner, to hold on by his fore paws until he scrambled his hind legs upon the sill, and then forced himself through an aperture so small that he could not jump, but had to let himself fall. The window is at a considerable height from the ground; and, as a rather wide trench runs around the building, Apollo had to make a tremendous leap to reach the window-sill. He had evidently failed several times, the scratches on the old wall showing where he had slid down. He always was a fine jumper, but this window must have tested his leaping powers to the utmost.

Sometimes we see in mankind an instance of good-hearted blundering, wrong-headed honesty; and much the same mixture of characteristics is to be found in the dog.

There was a brilliant black-and-tan terrier, named "Boxer," belonging to a Mr. B——, who was then in India, and about to proceed on the welcome journey home. Boxer had one prevailing idea in his doggish mind, namely, that he had perpetually to take care of some one or something. He watched his master's property with the utmost fidelity. Once, after the return of the family to Scotland, a couple of Irish beggars came by, and were given a good meal, the empty dish to be left outside the house. When they had finished, the woman, seeing that the cook was not in the kitchen, slipped in and stole a loaf of bread. She had not calculated on Boxer, who was out in a moment, caught the woman by the bare ankle, and there held her until his master came himself to take charge of the thief.

Had the dog restricted himself to such guardianship, he would have been a most excellent guardian; but, unfortunately, he was possessed with a rooted idea that every one who approached his mistress meant to hurt her, and must therefore be assaulted. When she was ill, and lying on a couch, he used to sit by her side, and was so careful in his watch that he would not allow even her husband to approach without seizing him. He did not hurt his master, though he bit his ankle a hundred times, by way of reminding him that his mistress was not to be disturbed.

In one way he was really useful, especially during the residence of the family in India. During her illness, his mistress had a very great antipathy to centipedes, cockroaches, and other creeping things, of which there is ample store in that country. Boxer somehow found out that they were obnoxious to his mistress, and used to keep a sharp lookout for them if they approached her. Sometimes, if he were not at hand, and he heard a scream, he would dash off to his mistress, look about for the cause of her annoyance, and straightway demolish it.

In his anxiety to do his duty to his mistress, Boxer sometimes allowed his zeal to outrun his discretion.

Once, during the voyage, the ship was becalmed in the tropics, so that the man at the wheel had a sinecure. Mrs. B—— was lying in the cabin at the time. The man, seeing a needle lying just outside the door of the cabin, went and picked it up, and was instantly pinned by Boxer, who chose to think that he was stealing the property of his mistress. He did not hurt the man, but frightened him so much that he hallooed loud enough to alarm all the inmates of the ship. Among others, Mrs. B—— ran out to see what was the matter, and advised the man to put the needle down again. This he did, when Boxer at once released him.

He behaved in a somewhat similar manner when the ship arrived in the Cove of Cork, though in this instance with more show of right. The stores of biscuit had been got up on the main-deck, for the purpose of ridding them of the cockroaches, weevils, and other unpleasant creatures that are apt to infest provisions. A number of Irishmen came on board with milk, eggs, etc., for sale, and one of them, thinking no harm, began to eat a biscuit. Boxer, however, considered himself the guardian of the ship's stores, flew at the man, and drove him away.

When home was reached, he took, in his wrong-headed way, a violent antipathy to the clergy-man. Perhaps he objected to a black dress, after being accustomed to the light costumes of India. At all events, he could not endure the gentleman, and always seemed to know instinctively whenever he was approaching the house. On these occasions it was necessary to shut him up; and even then he used to tear and scratch at the door so furiously that he greatly damaged it. The oddest part of the proceeding was that as soon as the gentleman was in the room with his master and mistress Boxer did not trouble himself about him.

This queer, faithful, blundering dog lived for nearly twenty years in the family.

A very common form of conscience among the lower animals is that which may be defined as a recognition of having done wrong, and an acknowledgment that punishment is deserved. It is exactly the same feeling which induced Adam to hide himself after he had fallen into sin. Animals have in their way very decided ideas as to right and wrong; and when they have committed an act which they know will offend their master, they display as keen a conscience as could be exhibited by any human being self-convicted of a sin; and, in many cases, the offense is acknowledged, and the creature remains miserable until pardon has been granted. This we call in ourselves penitence.

Two examples of this phase of conscience are here given. As to the first, I was in doubt whether to place it under the head of Reasoning, Language, or Love of Owner. But, as it illustrates the power of conscience in the lower animals, I have placed it under the present head, without, however, removing the passages relating to the other qualities:

"Reasoning powers are certainly exercised by dogs; how would they otherwise know when Sunday came around? Our large dog, 'Bran,' a cross between a retriever and a deerhound, never thinks of following us to church, though he regularly comes in on Sunday afternoon, in expectation of the walk which he knows his young masters take between the services; and on week-days he will even run up-stairs if he hears us moving about the bedrooms, which he in some way connects with walking out. He looks so intelligent that it is difficult to believe he does not understand conversations, and we talk to him often as if he were a human being. He is very good-tempered, and particularly so with cats and children. When we were at Worthing, two years ago, a large white cat belonging to the house constantly shared his bed; and on more than one occasion the cat, dog, and the little grandchild of our landlady were found curled up together.

"Whenever he did wrong as a young dog, we found the greatest punishment was to take no notice of him, and refuse his offered paw. On one occasion, I remember, he ran off, and was missing all day. When he came back, he was shut up in his sleeping-place, after we had shaken our heads at him and turned away. Although he must have been very hungry, he would not touch his food, but sat close to the door, whining and crying, till we made it up with him by telling him he was forgiven, and taking his offered paw, when he ate his supper and went quietly to bed. His love for us is unbounded, and he almost overwhelms us sometimes by his affectionate embraces, especially if we have been away, when he almost talks in his joy at seeing us again."

A lady has sent me a short account of the behavior of another dog, which clearly shows that the animal possesses the attribute of conscience. The little animal has been taught many tricks, among which is the accomplishment of shaking hands. This he will seldom condescend to do without much coaxing; but if he has done anything wrong, he comes up, looking very much ashamed of himself, and voluntarily offers his paw.

I may here refer to the dog "Help," who went sheep-killing while his master thought that he was chained up at home. It was a clear case of conscience, though not accompanied by penitence. He knew that he was acting wrongly, and that his master would be offended, and therefore endeavored to avoid punishment by destroying the evidence of his crime.

How painfully keen can be the sense of conscience in the dog is shown by the following account, which is written by a brother clergyman well known in the literary world:

"A Newfoundland dog of great age, but still the gentle, good-tempered friend of his master's children, lay one morning sound asleep. One of his playmates, wishing that he should accompany their walk, gave him a kick. The poor dog, suddenly awakened, seized sharply the little girl's leg, but without really hurting her. The nurse thereupon scolded him, pretended to beat him with a pocket-handkerchief, and, when he wanted to go with them, shut the door in his face.

"One of the men soon afterward found him lying with his head in a ditch, dragged him out, and brought him to the stables, where he lay, refusing to eat or drink. Ere long he was again found at the same ditch, dead. Whether, in remorse and despair of forgiveness, he had successfully repeated an attempt at suicide, or whether he had lain down there to die of a broken heart, I do not know."

One or the other was evidently the case, and, whether it were suicide or sorrow, conscience was the real cause of death.

The same writer proceeds to say: "You also asked for the epitaph on our poor little friend's grave. It is as follows:

'"COLL,"

'FAITHFUL, LOVING, GENTLE, WISE,
BY HIS UNTIMELY DEATH
MADE EMPTY NO SMALL SPACE
IN OUR HOME AND HEART.

'Alas! too soon, dear loving friend,
 Our close companionship doth end;
 Yet sense of Right, heart true and fond,
 Must have, methinks, some glad BEYOND.'

"Poor H.! it was her first great grief, and yet lives. A cousin spoke lightly of the epitaph as she stood by the grave the other day in her garden.

"'Please come away, G.,' said H., 'and don't let us speak about it. Something has been left out in your composition; *you* can not understand.'"

With regard to the supposition that the former of these dogs committed suicide, it is not so groundless as might be supposed. Dogs certainly know that water will drown other beings, or they would not take the trouble of rescuing them; and it is therefore but natural to infer that they are aware that the same element will drown themselves. There is more than one instance known of a dog deliberately drowning itself; and the instance which has just been narrated looks very much as if the same course had been adopted.

The following little story is one of Lady E.'s anecdotes, and shows how not only dogs, but cats, can possess a sense of moral responsibility:

"I trust the following anecdote of my cat 'Rosy' may be found interesting.

"You know that she was given to me when quite a kitten, and she is now nearly fourteen years old. She has always had a great aversion to dogs, and, no matter what their size might be, she would drive them away if they came on our premises.

"Whenever the servants left the kitchen, she would sit near the door, and, if a stranger approached, growled like a dog. One day the cook had left the cat alone, and the butcher's boy came for orders as usual. Receiving no reply, he opened the door and walked inside. Perceiving him to be a stranger, Rosy, to his surprise, flew at him, and held him tightly till the cook returned. Instead of being angry at the attack, the lad admired her bravery, said she was as good as a house-dog, and often rewarded her with meat from his shop."

In Bennett's "Wanderings in New South Wales," the Siamese ape "Ungka" is mentioned as possessing the sense of moral responsibility, though the mode in which it was manifested was rather of the ludicrous than the lofty kind:

"One instance of a very close approximation to, if it may not be considered absolutely an exercise of the reasoning faculty, occurred in this animal.

"Once or twice I lectured him for taking away my soap continually from the washing-place, which he would remove for his amusement from that place, and leave it about the cabin. One morning I was writing, the ape being present in the cabin, when, casting my eyes toward him, I saw the little fellow taking the soap. I watched him, without his perceiving that I did so; and he would occasionally cast a furtive glance toward the place where I sat. I pretended to write; he, seeing me busily occupied, took the soap, and moved away with it in his paws. When he had walked half the length of the cabin, I spoke quietly, without frightening him. The instant he found I saw him he walked back again, and deposited the soap nearly in the same place from which he had taken it.

"There was certainly something more than instinct in that action. He evidently betrayed a consciousness of having done wrong, both by his first and last actions; and what is reason, if that is not an exercise of it?"

I know a little child, not two years old, whose favorite amusement is to get at a box full of Windsor soap, and disperse the cakes all over the room, in all sorts of places. She is not allowed to do so without permission; and more than once, when she has been detected in doing so, she has acted exactly as Ungka did, *i. e.*, replaced the soap, and tried to look as if she had not touched it. In both cases the process of reasoning is identical; and so is the sense of conscience, or moral responsibility.

A curious example of the power of conscience is related by Mr. Mansfield Parkyns, in his well-known work on Abyssinia. He had a semi-tamed hunting-dog (one of the wild animals of the country), and was much interested in the habits of the animal, which he named "Tokla."

"Once I remember being attracted into the yard by a bustling noise as of animals running about, intermixed with my pet's shrill, squeaking voice. On going out, nothing was apparent but a sheep lately bought for dinner, which, however, was running about with every appearance of nervousness. There was Tokla, whose voice I had just heard uttering notes of unusual excitement, lying quietly in a corner, shamming sleep, but peeping at me from a corner of one of his little wicked black eyes.

"I said nothing, but concealed myself in a shed, through the branches that formed the sides of which I could observe all that passed. For a short time the little brute lay motionless in the same position as I had left him. After a while, however, he got up stealthily, stretching himself as if just awake, but at the same time taking a furtive glance to see that all was quiet. Having satisfied himself on this point, he made a rush at the poor sheep, with his ears back, and squeaking horribly. The sheep ran away when it could, only standing and butting at its little opponent when driven into a corner, and evidently in a desperate fright.

"Tokla seemed to heed little whether hoof or horns met his advances, but kept on, now rushing furiously in, now dodging for a more favorable opening, incessantly for half an hour. I doubt not, though scarcely six pounds' weight, he would have ended by walking into the mutton of his adversary had I not felt compassion for the poor sheep's sufferings, and disturbed my little friend in his pursuit. Indeed, I could not have allowed him to indulge his sporting propensities so long as he did except as a study of his natural ideas, manners, and customs."

Here is a distinct case of conscience, and of cheating in order to conceal his delinquency. He was perfectly aware that he was doing wrong in attacking the sheep, and so feigned to be sleeping when his master came on the scene. This is the more curious, because Tokla was not one of the domestic dogs, but a predacious animal which had only been recently and partially tamed.

Almost every one who has possessed pet animals must have noticed how often they exhibit remorse, *i. e.*, a keen sense of having done wrong, their conscience having convicted them of their misconduct, and their whole demeanor showing that they are sensible of their fault.

Here is an example: A Mr. B—— had a magnificent staghound, named "Gwynne." The dog had one fault: he was not fond of children, and therefore was given away, as unsuitable for his owner's house.

His new master lived in Sutherlandshire, and sent the dog to one of his farthest farms, where he was taken in charge by the shepherd's wife. One morning, after the woman had made the porridge for breakfast, she went out of the house, and on returning met the dog, who had evidently been eating the porridge. With an expression of anger she struck him on the head, whereupon Gwynne left the place, and was never seen again, though advertisements and liberal offers of reward were issued.

This, however, is not all. Several times previously he had been given away, and had always made his way back to his old master; but this time he did not do so, evidently because he felt himself rightfully in disgrace for bad conduct, and he did not dare to show himself in his master's presence.

CHAPTER XIII.

SYMPATHY AND FRIENDSHIP.

Love and its Various Phases of Development.—Sympathy between Animals of the Same Species.—Dr. J. Brown's Story of "Nipper" and the Destitute Pointer.—Protection of the Weak.—"Pizarro" and the Terrier.—"Mungo" and his Big Friend.—The Ludicrous Side of Sympathy.—Church Bells and their Effects.—Cats and their Comrades.—Division of Labor.—Sympathy in the Weasel.—Sympathy between Animals of Different Species.—Several Anecdotes of Dogs and Cats.—The Dog and the Persecuted Cock.—Dog Feeding Kids.—Cat-sympathy.—The Grandmother, the Daughter, and the Dead Grandchild.—A Generous Redbreast.—Animals Sympathizing with Man.—Story of "Nelly" and her Mistress.—Nelly's Death and Last Request.—"Prettina's" Sympathy with a Sufferer.—"Flo," the Family Consoler.—Friendship Defined.—"Pincher" and the Quarrelsome Sheep-dog.—Friendship between Cats.—The Story of "Dick" and "Kate."—Kittens and Dog.—Friendship between Cows and a Sheep; Cats and Horses.—Friendship between a Horse, a Cat, and a Lame Chicken.—Friendship between a Cat and Ducks.—Friendship between a Dog and Ducks; Java Sparrows and Doves; a Monkey and a Hen.

WE are now approaching the loftiest characteristic which adorns humanity—namely, Love—and are about to inquire how far it is shared by the lower animals. It has many phases of development, the first of which is sympathy, *i. e.*, the capacity of feeling for the sufferings of another. I shall show that many and perhaps all living creatures possess the capacity of sympathy, and that in numerous cases it is not restricted to their own species, but is extended to those beings which appear to have very little in common with each other.

Usually, however, sympathy is exhibited between animals of the same species, and is often seen in the dog. Such, for example, is the well-known instance where one dog was seen supporting the broken leg of another; also the fact that a dog which has been cured of some injury will take a fellow-sufferer to his benefactor—an example of which I knew personally. I need hardly observe that such sympathy could not be carried out unless the animals possessed a language sufficiently defined to transmit ideas from one to the other.

I will begin with a few instances of sympathy between animals of the same species, and place at their head Dr. J. Brown's graphic account of his dog "Nipper:"

"Many years ago I got a proof of the unseen and therefore unhelped miseries of the homeless dog. I was walking down Duke Street, when I felt myself gently nipped in the leg. I turned, and there was a ragged little terrier crouching and abasing himself utterly, as if asking pardon for what he had done. He then stood up on end, and begged as only these coaxing little ruffians can.

"Being in a hurry, I curtly praised his performance with 'Good dog!' clapped his dirty sides, and, turning around, made down the hill; when presently the same nip, perhaps a little nippier—the same scene, only more intense—the same begging and urgent motioning of his short, shaggy paws. 'There's meaning in this,' said I to myself, and looked at him keenly and differently. He seemed to twig at once, and, with a shrill cry, was off much faster than I could follow. He stopped every now and then to see that I followed, and, by way of putting off the time and urging me, got up on the aforesaid portion of his body, and when I came up was off again.

"This continued till, after going through sundry streets and by-lanes, we came to a gate, under which my short-legged friend disappeared. Of course I couldn't follow him. This astonished him greatly. He came out to me, and as much as said, 'Why don't you come in?' I tried to open it, but in vain. My friend vanished, and was silent. I was losing in despair and disgust, when I heard his muffled, ecstatic yelp far off around the end of the wall; and there he was, wild with excitement. I followed, and came to a place where, with a somewhat burglarious ingenuity, I got myself squeezed into a deserted coach-yard, lying all rude and waste.

"My peremptory small friend went under a shed, and disappeared in a twinkling through the window of an old coach body, which had long ago parted from its wheels and become sedentary. I remember the arms of the Fife family were on its panel; and I dare say this chariot, with its C springs, had figured in 1822 at the King's visit, when all Scotland was somewhat Fifeish. I looked in, and there was a pointer bitch, with a litter of five pups; the mother like a ghost, and wild with maternity and hunger; her raging, yelling brood tearing away at her dry dugs.

"I never saw a more affecting or more miserable scene than that family inside the coach. The poor bewildered mother, I found, had been lost by some sportsman returning south, and must have slunk away there into that deserted place when her pangs (for she has her pangs as well as a duchess) came; and there, in that forlorn retreat, had she been with them, rushing out to grab any chance garbage, running back fiercely to them—this going on day after day, night after night. What the relief was when we got her well fed and cared for—and her children, filled and silent, all cuddling about her asleep, and she asleep too—awaking up to assure herself that this was all true, and that there they were, all the five, each as plump as a plum—

"All too happy in the treasure
Of her own exceeding pleasure;"

what this is in kind, and all the greater in amount as many outnumber one, may be the relief, the happiness, the charity experienced and exercised in a homely, well-regulated *Dog Home*.

"*Nipper*—for he was a waif—I took home that night, and gave him his name. He lived a merry life with me—showed much pluck and zeal in the killing of rats, and incontinently slew a cat which had—unnatural brute—unlike his friend—deserted her kittens, and was howling offensively inside his kennel. He died, aged sixteen, healthy, lean, and happy to the last. As for *Perdita* and her pups, they brought large prices, the late Andrew Buchanan, of Coltbridge, an excellent authority and man—the honestest 'dog-man' I ever knew—having discovered that their blood and her culture were of the best.

"I have subscribed to the London 'Home' ever since I knew of it, and will be glad to do as much more for one of our own, as Edinburgh is nearer and dearer than the city of millions of dogs and men. And let us remember that our own dogs are in danger of being infected by all the dog-diseases, from the tragic *rabies* down to the mange and bad manners, by these pariah dogs; for you know among dogs there is in practical operation that absolute equality and fraternity which has only been as yet talked of and shot at by and for us."

In this charmingly told anecdote, we see not only sympathy, but self-denial, reasoning, and a power of communicating ideas to a human being. Being a waif and a stray himself, without a master, and dependent upon chance for food, the little animal took compassion on his suffering companion, and went out to beg from man the assistance which he was not able to render himself. But that assistance was not meant for himself, however much he needed it, but for his companion who needed it more. Doubtless his instinct, and not his reason, taught him to select the person to whom he applied; for it is not every man who will allow himself to be nipped in the leg without repaying the bite, however gentle, with a kick or a blow. Animals, like children, always know their friends.

See, for example, when he forgot to calculate the difference of size between his newly found friend and himself. Finding that the man could not crawl under the gate like himself, the dog calculated the dimensions of the man, and pointed out to him an aperture through which he could make his way.

A lady writes to me to say that a friend of hers has two dogs—one a Newfoundland, and the other a small black-and-tan terrier. They are both good water-dogs, and are now in the habit of swimming about together. But, on the first occasion after their introduction to each other, when the terrier jumped into the water, the Newfoundland dog sprang in after him and put him on the bank, evidently thinking that he had fallen accidentally into the water and might be drowned.

The following story is sent by another lady:

"We had a noble blood-hound, 'Pizarro,' sent us from Manilla; although his kind are supposed to be more or less savage, he was most gentle when well treated. When on board ship, he became much attached to a small terrier, which no one dared to molest in his presence. The sailors used to take the little terrier in the boat with them, leave Pizarro on board the ship, and commence teasing his little friend—a proceeding which rendered the blood-hound most irritable. He afterward extended his protecting power to one of those ornamental but useless Danish dogs, which we had at that time, and who always rushed to his kennel for protection if threatened with punishment, and then no one dared interfere."

Several anecdotes, of a character somewhat similar to the following, are tolerable well known. I am glad, therefore, to present my readers with a story which possesses the double advantage of being both original and authenticated:

"In a little village in Wiltshire there lived a small black terrier, called 'Mungo,' and a large yard-dog, the two being on the most amicable terms. One night the terrier paid a neighboring farm-house a visit, in order to offer his respects to another little terrier, whom he much admired. But, alas for his gallant intentions! a large rough watch-dog, not tolerating rivals, set on him savagely, and poor Mungo returned home in a sorry plight—bleeding, torn, limping, and scarcely able to crawl.

"He lay down by his faithful friend, and told of his piteous wooing. Fondly and gently the big dog listened, and licked his friend's wounds, who for many a day lay sorely bruised, and never attempted to leave home. Some time afterward, on a fine moonlight night, some laborers, who were returning home across some fields, met the two friends trotting gayly along. Next morning the farmer found his savage watch-dog stretched stiff and stark on the straw in his yard."

What a combination of qualities do we not find in the conduct of these two dogs. They must have possessed a language sufficiently definite for the one to tell the other what had befallen him, and to designate the offender. They then must have arranged that the big dog was to avenge the injuries inflicted on his little friend as soon as the latter was well enough to show him the way. There was memory in both dogs, enabling them to postpone the execution of their design until the injured dog had recovered; and there was sympathy for suffering in the large dog, and desire for revenge in the little one. The two dogs in question belonged to a clergyman, who told the story to my correspondent.

There are instances among mankind where even the best of feelings present a ludicrous side, and animals are not exempt from this rule. In the following example the sympathy was very well meant, though the mode of showing it was exceedingly ludicrous:

"I have said that my dog 'Lion,' a cross between the setter and the sheep-dog, always sets up a piteous howling when the neighboring church bells begin calling the villagers to morning service at eleven. At the first slow tolling he takes no notice of the bell, but as they ring the changes he becomes uneasy and wanders about; and when the chimes begin, he no longer contains his feelings, but howls with all his might. He is by no means a howling dog, and bears all the other ills of life with patience, or, at most, indulges in a whine.

"Now my mother's dog 'Snap,' a pure Skye terrier, is, as a rule, supremely indifferent to bells, though she will bark by the hour together at a treed squirrel or at some distant or even imaginary sound. No sooner, however, does she, when on a visit to my cottage, hear Lion's cry of distress and remonstrance, than she joins the chorus, and the two dogs will sit on their haunches, with nose in air, and howl there until I call them in, or until the bells abate their noise. The one dog affords a good picture of sorrow, and the other of sympathy."

Perhaps Snap thought that it was only good manners to show that she felt for Lion's sufferings, though she did not share them herself, and so she joined him in his lamentations.

It may be, however, that Lion had more cause for complaint than we might fancy. There are many dogs to whom certain sounds are not only obnoxious, but actually injurious. This is well corroborated by a curious anecdote communicated to me by the late J. Hatton, M.D.:

"With regard to the effects of sound upon animals, I remember, when a child, my brother going into the country, on a Sunday afternoon, to see a friend, who gave him a young pup, about ten weeks old. At that time we lived in the precincts of the Cathedral. The ringers were accustomed to practice every Sunday evening at eight o'clock. No sooner did they begin than the dog began to run round and round the room at a furious rate, and finally rushed under the sofa, where the poor animal almost immediately died in convulsions."

Cats are often kind to each other, sympathizing under difficulties, and helping their friends who need assistance. One of my friends is a great admirer of cats and their disposition, and has noted many of their ways. One of her cats was rather a weak animal, and was unable to carry her kittens about after the manner of cats. So, when she wished to carry her kittens from one place to another, she was accustomed to impress a stronger cat into her service, she walking by the side of her friend in order to act as guide.

Another of the cats, when oppressed with the cares of a family, did exactly what a human mother does when she can afford it. She employed a nursemaid, *i. e.*, she brought a half-grown kitten, and placed it in charge of her young while she went for a ramble.

In Hardwicke's *Science Gossip*, which is really a treasure-house of information for those who know how to use it, are many anecdotes of sympathy between animals. One of these anecdotes shows reason as well as sympathy in the common weasel.

A clergyman was driving along the road near Basingstoke, when his horse trod on a weasel, which could not get out of the way in time. The little animal was paralyzed, its spine seeming to be broken, so that it could not move its hind legs. Presently another weasel came out of the roadside, went up to the injured animal, and, after carefully inspecting the invalid, picked it up and carried it to the side of the road, where it would not be endangered by traffic.

Another case of sympathy between creatures of the same species is given in the same journal. A female wood-pigeon was sitting on her eggs, and her mate was close at hand. A heavy shower of rain came suddenly on, whereupon the male bird took up a position above his mate (who could not leave her eggs), and with his spread wings formed a shelter from the rain.

We will now pass to sympathy between animals of different species. The hereditary enmity between cat and dog is proverbial; and yet, when in good hands, they are sure to become very loving friends, and even to show considerable sympathy with each other. Here is a case in which the animals had but the slightest passing acquaintance with each other. The anecdote which immediately follows is communicated by the same lady:

"Compassion was shown in the following case. A poor little cat was lying very ill by the kitchen fire; another cat came inquiringly up. A Scotch terrier (belonging to the house in which we were then lodging, and therefore a comparative stranger to the invalid) immediately jumped off a chair, and silently but firmly turned it back, as if to say, 'You must not disturb her.' He also turned back in the same manner our own large dog."

"Many years ago my mother had a cat and dog; and when the cat had kittens, the dog, a terrier, would take charge of them for an hour at a time, and no one dared touch them. Although at other times he was gentle, he then snarled at all comers. Directly the mother reappeared, 'Fly' walked off and resigned his charge."

The four following anecdotes all relate to sympathy between cats and dogs, and have been sent by different correspondents:

"One day a large black cat entered the garden in a most deplorable condition, her tail nearly cut in two by a tin kettle which had been tied to it. The kettle was taken off, and the poor creature brought into the house and fed. Our little dog 'Trotty' was greatly delighted to have another friend; but 'Blackie,' as we called the cat, would not allow him to go near her, scratching and spitting if he approached. All the time her tail seemed very painful, but at the end of three or four days Trotty somehow managed to bite off the end of it. This eased the poor creature's pains, and from that time they were loving friends."

Trotty's reasoning was as correct in this case as if he had been the subject of transmigration, and had formerly inhabited the body of a hospital surgeon.

"When our little dog Trotty was quite young, we had a kitten, 'Mittie' by name. She was a gentle, loving creature, who evidently disliked being pulled about and teased by Trotty, but only resented it in the most gentle way. The result of this teasing was that poor little Mittie did not grow, and at ten months old she was a dwarf.

"She was then accidentally scalded, and so badly that for some weeks she lay on a pillow, and had her sores regularly dressed with oil. All this while Trotty was apparently troubled, and when the sores were partially healed he gently licked them, and so aided in her recovery. From that time he never teased her, and they lived together for a year, Mittie growing into a fine cat.

"At the end of this time I fancied that Trotty was again at his tricks, but on closely watching them I found that Mittie held up her head in order that Trotty might lick her neck, on which we found a small lump. This went on for some time, Mittie touching Trotty with her paw when she wished to be licked, and again when she wished him to desist. The lump proved malignant, and dear little Mittie died. Trotty was restless for weeks, and would not eat as usual."

"'There was a ferocious bull-dog kept as a guardian of the premises. He was so fierce that on one occasion he tried to bite his mistress because she ventured too near his kennel. Once, however, he showed that he was not deficient in kindly feelings, and that they might have been developed by proper management.

"One day a little kitten got out of a window three stories high, and fell on the stone paving of the yard near the dog's kennel. It was so hurt and crushed that even its mother would not go near it. The dog, however, picked it up carefully, took it into his kennel, licked it clean, and nursed it carefully till the poor little thing died."

"'Bandy,' our turnspit dog, was an inmate

of the house before Miss 'Chinchilla' Puss came to reside in it ten months ago. She was very wild and frightened when she came, dogs being her natural abhorrence; but by constant association at the dinner-table, round the fireplace, and even in the ladies' rooms, they became good friends; and their love was cemented on the occasion of poor Bandy having had a fit and being unable to move for some hours. Miss Chinchilla came frequently to inquire for him, and greeted him with a kiss — literally they touched lips and noses.

"Time passed on, and Chinchilla arrived at the dignified state of married life, and a month ago she had her first kitten. On the occasion, Bandy's anxious solicitations were constant and gentle, nor could his curiosity be satisfied on hearing the kitten's wee voice till he had seen and smelt Kitty. The sight of such a warm small lump of life excited an interest which was far greater than curiosity, and whenever Chinchilla left her kitten for any time, he took upon himself the office of guardian and nurse, for he licked and watched it as if it had been one of his own offspring. The other evening, while the mother was away, he even got into the cat's basket, and curled himself up in it, as she does, while the kitten lay upon him and under his paw exactly as it does with its mother. Now at a month old Kitty goes up to him and 'shows fight' (as children say when they want to have a stand-up game), while the mother looks on, waiting to share in the fun.

"I am hoping to see, when Kitty grows up, that she will have extinguished the natural antipathy of her race to the dog tribe; if so, I think it will be a beautiful proof of cultivation and domestication obliterating the prejudices of instinct and hereditary habits. Since it is the affection and instinct that form the life of all existences, I believe and hope we are cultivating the immortal principle in our domestic animals by subduing the more animal nature, and educating their affections and intelligences, and so developing a higher race of dogs and cats, or horses, or any other pet.

"I should like to know if you think that in the next stage of existence our animals will know us individually as they do now in a measure by smell and scent and voice. I am anxious to see your work on the proofs of immortality in animals. I do so hope that, when I pass beyond the veil, I shall know and be recognized by a dear old pet dog, 'Beppo,' a most devoted animal, who lived and died with us."

A gentleman living in Edinburgh has just sent me this remarkable anecdote of sympathy in a dog, showing how wide can be a dog's sympathies, and how cleverly he can carry them into effect:

"I once gave a spaniel, called 'Jack,' to a farmer friend in the neighborhood of the city. Jack's kennel was placed in the farm-yard, where the poultry were daily fed. Among them happened to be a poor, unfortunate, unpopular cock, which was not allowed to have a share of what was going, but was punished severely whenever he made an attempt to get any food.

"Jack somehow observed this, and, feeling sympathy for the poor bird, was seen daily to leave some of his food, to carry the 'bicker' which contained it into his kennel, and wait there until all the poultry were gone. He would then take his bicker outside, put it down where the cock could get it, and stand on watch all the time in order to protect him. Sometimes he would leave the bicker inside the kennel, and, if the bird were near at hand, he would go round about him until he got him into the kennel, so that he might take his food without being disturbed.

"I regret to say that Jack is now dead; but he was a dog of more than ordinary canine parts for he exhibited a sagacity and sympathy toward that sadly tormented bird which showed that he was an animal of a rare stamp, and far above his fellows."

I hope that the reader will appreciate the character of this dog as it deserves, and see how he displayed virtues of which any human being might be proud. There is compassion for sufferings unjustly inflicted upon a fellow-being, and a determination to redress them. Then there is self-denial in depriving himself of his food, generosity in giving it away, and high reasoning power as exemplified by the various means employed in managing that the poor persecuted bird should have its food in peace, undisturbed by its heartless fellows.

The end of this strange friendship was very remarkable.

The ill-usage of the other birds still continued, and at last the cock was accustomed regularly to take refuge in the dog's kennel. Probably from the perpetual bullying which he endured, he fell ill, and one morning was found dead in the kennel, lying closely pressed to his only friend.

Giving food, indeed, seems to be a favorite way by which animals of different species express their sympathies with each other. The following little history is given in the *Zoologist*, page 9649 :

"Last year, when the troops left this station to proceed to the frontier war, a goat belonging to an officer had two young kids the very morning the force marched. The cruel native servants, who have less feeling than any animal, even a tiger, took with them the poor mother and left the two kids behind, because to carry them would have entailed a little trouble — a thing most devoutly abhorred by this class of menials.

"The little kids made a terrible bleating noise at being left all alone; and a pariah dog, who was employed as a wet-nurse in the opposite compound for two English puppies, came over the road and took the helpless little kids in her mouth, and conveyed them to the box where her two puppies were. After this she regularly suckled them, and brought them up with the other two of her adopted family. It was a curious sight, the old lady suckling two puppies and two kids. She lay down to the former, but had to stand up for the latter, for they used to run at her in the usual vehement way lambs and kids do at their mothers, which often gave the dog great pain; but notwithstanding this she was never known to bite at them.

"These two kids grew up and followed the dog about, along with the puppies, all day, until the kids became as big as the old dog herself; she nursed them for about three months, when she had a family of her own, and left off taking any notice of them further than by a good-humored wag of the tail or an occasional lick of their faces. These kids grew up to be big goats, and continued playing with the dogs, their foster brother and sister. The old dog had been in the first place deprived of her own offspring, and the two puppies had been brought to her to bring up. Perhaps having lost her own family made her take compassion on the kids, thus showing that 'a kindred feeling makes us wondrous kind' does not apply to the human race only."

A curious instance of sympathy is related by Lady E.:

"Rosy's daughter was a curious tortoise-shell, with four white paws, which were always kept particularly clean.

"When she was nearly twelve months old I went into the room to breakfast, and perceived both Rosy and Tiney outside the window, resting their front paws on the window-frame, and carrying a dead kitten in their mouths, Rosy holding the kitten of a few hours old by the back of the neck, cat-fashion, and Tiney supporting the hind legs. Both mother and daughter were looking very solemn, as though soliciting sympathy for the death of Tiney's first-born child.

"Rosy has had a great many kittens since then, but Tiney never had another, though she has been most anxious and attentive to her little brother and sister kittens; and whenever Rosy left them longer than Tiney considered prudent, she would call her and drive her to her baby kittens, giving her an unmistakable box on the ear and a scolding for her neglect of her young ones.

"When they grew too large and heavy for Rosy to carry alone, the pair always bore the burden between them, Rosy taking them by the neck, and Tiney the hind legs, as in the first instance. In this way they frequently mounted the stairs, and it was extraordinary to see how well they managed together.

"Poor Tiney, who was a wonderful cat, and did most clever things, was lost while I was away from home for some months. Rosy has traveled about with me, and sits as quietly on my knee as a child would do. She likes to look out of the carriage window, and when any thing passing takes her fancy she puts her paw on my chest, and makes a pretty little noise, as though asking if I had seen it also."

There are many examples known of birds feeling sympathy with the lost or deserted young of other species, and taking upon themselves the task of feeding the starving children. The following case is really a remarkable one, for it is scarcely possible for two birds to be more unlike in their manners than the starling and the redbreast, the former being essentially a social bird, and the latter as essentially a solitary one, isolating himself with the greatest care, and always appropriating to himself some district which he is pleased to consider as his own property, and in which he will not allow another redbreast to show himself. Indeed, he does not like a bird of any kind to intrude upon his premises, and whenever they show themselves they must be very strong birds indeed not to be attacked by this jealous defender of his rights. The following little history is taken from Hardwicke's *Science Gossip* for September, 1871:

"A little redbreast has come to our door all through the winter for his meals, and a most friendly, welcome guest he has been. One spring morning we saw robin do a deed of charity that more than ever endeared the little bird to our hearts. It had been a bitterly cold night, and on our servant going down-stairs to fetch some coal to light the fires, she found a poor little starling shivering and frightened in the cellar. She called me to see the bird: it had only just left the nest, and it was so weak that it could not fly. I tried to coax it to eat, took it near the fire, offered it bread-crumbs, seeds, water; but no—the starling would not be tempted.

"Breakfast-time came, and with it the little robin. We thought that if we put the wee birdie out of doors its mother might come to look for her lost child; then came the fear of robin—he was so very pugnacious. Well, we risked it, keeping a very strict watch over the starling's safety. Robin eyed it for a moment, and flew away; still the little baby bird stood on one leg shivering, and no mother arrived. The moments seemed hours. Presently robin came flying back, and with something in his beak, too. Hop, hop, he came to where the wee baby starling was shivering, and popped a worm in its beak, which it opened, just as if robin had said, 'Open your mouth—here is some breakfast;' and away he flew, and again returned with some food to the young bird, and then they both flew away. We never saw the starling again, but good little robin's deed made him more loved than ever in the house."

I am rather glad to have the opportunity of making known these examples of sympathy between animals, because I have received communications from persons who really appreciate the moral capacities of the lower animals, but who can not bring themselves to believe that they feel any sympathy with each other, though they do so for man.

We now pass to another branch of the subject, namely, the capacity of the lower animals to sympathize with human beings in distress. The following touching narrative is from the pen of a lady:

"Some years ago we possessed a large watch-dog, a mastiff, who, when he became old, was allowed the free range of the garden.

"We also had a little Skye terrier, whom he took into his especial charge, walking with her, and apparently showing her the various walks, flower-beds, etc. She had, unfortunately, one great fault, *i. e.*, chasing the cat, who was also a pet. On one occasion she was taken in the act, and her master was administering a little castigation; whereupon the mastiff came up quietly to his master, and took his right arm in his mouth, not offering to bite, but asking him to withhold the coming stroke.

"The successor to this dog was a still more remarkable animal, belonging to the St. Bernard breed, named 'Nelly.' She came to us when six weeks old, and died in November, 1862, lamented not only by the household of which she formed a part, but by the whole neighborhood. Even strangers could not but notice her, for her face was full of soul, nobility, intelligence, and love.

"She was with us during a season of bitter bereavement. Her own altered looks, her quiet and sad demeanor, told how truly she shared in the prevailing sorrow. For many weeks she never entered the house (except the kitchen), but would often look wistfully up to the windows. At length, when she did venture into the dining-room, she merely walked direct to the well-known chair, and, finding it vacant, with saddened look turned away and left the room.

"As time rolled on, her visits to the house were renewed, and then it was that her sympathetic qualities were so touchingly displayed. She seemed to realize the change that had passed over us. She noticed our indications of sorrow when we thought that she was sleeping, and, leaving the spot where she was lying, she would offer us her paw with an expression of countenance which made itself felt.

"On more than one occasion she rose spontaneously from the warm rug, and, with a look which conveyed as impressively as words could do the sympathy which she felt, she rested her beautiful flaxen breast on the lap of the lonely one, clasped her in her arms, and licked the tear-bedewed cheeks."

The last scene of Nelly's life was very remarkable, as showing the complete understanding and sympathy which can exist between man and the lower animals. She entertained the profoundest affection for the old gardener, affection which was perfectly reciprocated.

"Her greatest trials were when George's duties called him away from her. At such times she used to station herself at the gate, eagerly listening for the coming footstep, with now and then a piteous howl. And when he did appear, what a rush of delight! what greetings! what fondlings!

"But I must hasten to the last sad scene. Our loving and much-loved Nelly died three days after giving birth to a litter of puppies. The best skill that was to be had was obtained, and her faithful George watched her by night and by day. With all a mother's forgetfulness of self, she fulfilled her maternal duties until the last day, when she evidently felt that she had nothing and could do nothing for them. She feebly rose from her couch, and gave her children into George's charge, with a look that said as plainly as human words could speak, 'Care for these helpless ones as you have cared for me.' What could a human mother do more? She then lay on the couch, from which her children had already been removed, stretched herself out, and, with her paws in the hand of her faithful friend, quietly breathed out her life.

"Can it be that virtues such as I have attempted to portray have found no fitting sphere for their exercise, but that, like the poor perished body, they have gone to destruction? Nay, even the body is not destroyed; it is only dissolved into other elements. Are we, then, to think that the all-wise Creator shows less respect for the immaterial than for the material; that while the inferior continues to exist, although in altered form, the superior is consigned to annihilation? Reason and analogy oppose it; revelation does not support it."

I have witnessed an example of sympathy with human sufferings shown by a cat. Her name was "Prettina," and she was grandmother of my own remarkable animal, who, although of a different sex, received the abbreviated name of "Pret," by way of honoring the memory of his beautiful grandmother. One day, while I was paying a visit, the cat's mistress was seized with a distressing cough, which used to last for a considerable time, and left her quite prostrate with the fatigue. As soon as the cough began the cat became uneasy, and at last jumped upon the couch on which her mistress was lying, uttered a series of sounds which evidently expressed pity, and laid her paw on the sufferer's lips.

This, I heard, was her invariable practice whenever a fit of coughing was prolonged more than usual.

As for sympathy displayed by dogs, there is no need for me to give examples. I suppose that no human being was ever free from troubles of some kind, and I am equally sure that no one who had a companionable dog felt that he was without sympathy. Doggie knows perfectly well when his master is suffering pain or sorrow, and his nose pushed into his master's hand, or laid lovingly on his knee, is a sign of sympathy which is worth having, though it only exists in the heart of a dog. From that moment a bond has been established between the soul of the man and that of the dog, and I can not believe that the bond can ever be severed by the death of the material body, whether of the man or of the animal.

I know a case where a dog was always the consoler. It belonged to a large family, and, as will be the case in families, one of the children occasionally got into disgrace, and was punished. Whenever this happened, "Flo" was sure to find out the sorrowing child, and, by licking its face and offering many caresses, would show her sympathy. One of the children was peculiarly sensitive, and, as if conscious that she specially needed sympathy, Flo would be more demonstrative toward her than toward the others.

FRIENDSHIP.

That friendship, which is another branch of love, exists among animals is a very well-known fact, exhibiting itself most frequently among domesticated animals. Horses, for example, which have been accustomed to draw the same carriage are usually sure to be great friends, and if one be exchanged the other is quite wretched for want of his companion, and seems unable to put any spirit into his work.

Dogs, too, are very apt to strike up friendship with each other, one or two examples of which I have already mentioned under other headings. One of my friends has a little terrier called "Pincher," who had in some way managed to make friends with a great sheep-dog. This was an unpleasant animal, of a quarrelsome disposition, and was always fighting some other dog. On these occasions Pincher always used to run to the assistance of his friend, and give him material help by attacking his adversary in the rear, snapping and barking, and biting his heels. This was very good of Pincher, but it was scarcely fair play to the other dog.

A very remarkable and affecting instance of friendship in a cat has been communicated to me by a lady:

"We had two kittens given us, fine, high-spirited animals, called 'Dick' and 'Kate.' They lived together happily for some time; but Kate was taken with fits, and, by the advice of the doctor, she was poisoned with prussic acid. She was buried far away from the house, because Dick was so fond of her that we feared he would find her grave.

"He did not see Kate removed, and of course had no knowledge of his loss except by his own instinct. But he hunted every where for her, called her in his way, and after the first day refused to eat. He went about the house in the most touching way, just like a person in grief, and at the end of three days he died."

The love which this affectionate creature bore to his companion was stronger than life, and I can not believe but that it survived death, and that the two loving creatures were again united in their own sphere of existence.

Among the animals, friendship is not confined to one species, but sometimes exhibits itself in animals which might be supposed to be peculiarly incongruous in their nature. Here, for example, is a case of friendship between a cat and a dog:

"A strong case of friendliness between cat and dog is to be witnessed in this house (near Guildford) at the present time. Two kittens, not related to each other, and about six weeks old, have been introduced into the house, at two different times, within the last seven or eight weeks. Instead of taking alarm at the sight of our big dog, they lick his face, bite his ears, and play with his tail; I believe they think that he wags it on purpose, and I am not sure but they are right; finally, when they are tired, they go to sleep beside him, so that it is not uncommon to find him with one kitten between his paws and the other leaning up against him; and if he walks about the house, one or both of the kittens will trot after him. Neither of these little creatures had seen a large dog until they came to us."

That cows and sheep live, as a rule, on good terms in the same pasture is a familiar fact, though sometimes the former are a little apt to bully the latter. I have, however, learned that a very strong affection can exist between animals so different, and that when accustomed to each other's society neither could be happy without the other.

"Some years ago we had a lamb whose mother died soon after its birth. It was brought up by hand on cow's milk, and, for the convenience of the feeder, was kept in the cow-house. It accompanied the cows to and from the field, and remained their companion for two or three years.

"The animal was quite a pet of the man who had charge of the cows, and he kept it with them until ordered by his master to place it with the sheep. After much demur, this was done; but for some days the man complained that 'Donald' was miserable, that he would not associate with the other sheep, and that they beat him. The master gave little heed to the statement, but at last gave permission for Donald to be restored to his old associates, and invited us to see the meeting.

"The cows all rushed to meet him, and he ran up to each in turn; but this was only a beginning. After a few minutes a cow went to Donald and began licking him from head to tail, and continued to do so until she had passed her tongue over every part of his body. He was then passed over to another, who did the same thing, until all the six cows had shown their affection."

That Donald should refuse to associate with the other sheep is not a matter of wonder, as he had been accustomed from his birth to the society of cows; but that the others should bully him is not so easily explained, except on the supposition that from long familiarity with cows he had contracted habits that were unsheeplike, and gave him a foreign air.

Horses are apt to contract friendship with different animals. The goat and the horse are frequent friends, and it often happens that a peculiarly vicious horse will allow a goat to take any liberties with him without dreaming of resenting them. The stable cat, too, is quite an institution in many places, the cat's usual place of repose being the back of the horse, and the horse being uneasy if left for any length of time without the society of his usual companion.

I know of one case where the friendship was exhibited in a very curious manner. A little kitten strayed, when very young, into the house of one of my friends, and was adopted by a cat who brought it up together with her own young. This kitten became a great frequenter of the stables, and made two rather odd friends, namely, a pony and a lame bantam. It was a curious sight to see the kitten and the bantam curled up asleep on the pony's broad back, where they would spend hours without being disturbed.

The horse and the goose have been known to be excellent friends for a long time, the bird rubbing his head, in the fondest manner, against that of the horse. I have mentioned, under another heading, the odd friendship that was struck up between a kitten and a brood of ducklings, the kitten always going to sleep on the ducks when they had settled down for the night.

Another odd instance of friendship occurred in the house of one of my friends.

He had a fine Newfoundland dog which took a fancy to a brood of young ducklings, and constituted himself their protector. They were quite willing to accept him in this capacity, and followed him about just as if he had been their mother. It was a specially interesting sight to watch the dog and the ducklings taking their *siesta*. The dog used to lie on his side, and the ducklings would nestle all about him.

There was one duckling in particular which invariably scrambled upon the dog's head, and sat on the eye which was uppermost, both parties appearing to be equally satisfied with this remarkable arrangement, though the dog must have been put to no small inconvenience by the pressure on his eye.

It is really curious to notice the apparently incongruous friendships which are often found among animals.

I knew of a monkey who was accustomed to live in a hen-house. He formed a friendship with one of the inmates, a hen which was in bad health, and the two were accustomed to sleep on the same perch, the monkey with his head nestled under the hen's wing.

I also know of two Java sparrows which always pass the night under the wings of two turtle-doves, which treat them like their own offspring.

CHAPTER XIV.

LOVE OF MASTER.

Attachment of Animals to Man.—Innate Yearning for Human Society.—"Jimmie," the Squirrel.—A Tame Sparrow and its Ways.—"Turey," the Rock-pigeon.—Sad End of a Pet.—Divided Allegiance.—The Dead Shepherd and his Dog.—Animals Dying for Love of Man.—"Phloss" and his Mistress.—My Dog "Rory." —My Children's Canary.—Mr. Webber's Account of the Bullfinch.—The Story of "Grayfriars' Bobby."— A Well-deserved Monument.—Power Possessed by Animals of Returning to their Masters.—A Collier Dog Finding his Way from Calcutta to Scotland.—"Zeno's" Singular Journey.—A Dog and his Complicated Journey from Manchester to Holywell.—Cats and their Supposed Attachment to Localities.—My Cat "Pret" and his Travels.—A Cat Crossing Scotland alone.—Suggested Source of the Power.—A Stray Persian Cat.—The Dog "Joey" and his Mistress's Letter.—The Indian Fakirs and their Tame Tigers.— "Rob," the Bloodhound, and the Child.—The Boy and the Savage Horse.—Two Pet Sheep.—Goose and "Goosey."—"Toby," the Gander.—Summary of the Subject.

UNDER this heading I place that feeling which induces animals to attach themselves to human beings, the feeling being the same whether the object of it be technically a master or not.

I have already referred to the intense yearning for human society which is felt by many of the lower animals, and which is indeed but the aspiration of the lower spirit developed by contact with the higher. In those animals which are domesticated, and therefore in perpetual contact with man from birth, this feeling is no matter of wonder. But that it should be exhibited in the non-domesticated animals and birds, and even, as we have already seen, in insects, is a fact which is well worth our consideration, as giving a clew to some of the many problems of life which are at present unsolved.

The power of attraction which is exercised by the spirit of man upon the spirit of the lower creation is well exemplified by the well-known fact that many of the wild animals will attach themselves to human beings, and will forsake the society of their own kind for that of the being whom they feel to be higher than themselves.

One of the wariest animals is the wild squirrel, as any one will say who has tried to approach one. He is horribly afraid of human beings, and if a man, woman, or child come to windward of a squirrel, the little animal is sure to scamper off at his best pace, scuttle up a tree, and hide himself behind some branch. Yet, as the following anecdote shows, the squirrel, wild as he may be, is peculiarly susceptible to the influence of the human spirit, and for the sake of human society will utterly abandon that of its own kind. The little history which is here given was sent to me expressly for this work:

"The squirrel was given to me while I was an undergraduate at Cambridge, in the summer of 1854. He was very young, and could scarcely jump from the table. I took him home with me in the long vacation, and he soon became so fond of me that when I went for a walk I used to take him with me, and he followed me like a dog.

"Although he had one of those whirlabout cages, it was with difficulty that I could keep him there, as when awake he preferred to follow his own devices, and, when tired, he usually slept on a soft cushion on the sofa; or, if the doors were left open, he would find his way into some bedroom, and nestle under the pillows.

"At night he always used to sleep with me, though he was rather troublesome, as nothing escaped his notice, and he always tried every thing with a nibble. He used to hide things dreadfully, and ladies' work-boxes were perfect mines to him. I am afraid he was rather encouraged in this, as my mother generally had a nut reserved in the corner of her box.

"In the morning, while I was dressing, I used to open my window, when 'Jimmie' used to get out, climb down a rose-tree that was nailed to the wall, and amuse himself by taking a run before breakfast. Afterward he usually went out again, and played about the lawn and plantation for three or four hours, returning by the window, and going to sleep on his favorite cushion.

"Once, when I was staying from home, and had taken Jimmie with me, I lost him for two nights. He had been playing about in the garden, and, being in a strange place, had evidently lost his way. I was very unhappy about him, and had given him up for lost, when on the second day I heard his feet pattering near the door, and joyfully welcomed him back. When I came over to Jersey, I brought my little friend with me; but in 1858 the poor little fellow caught cold, became paralyzed, and soon died, to my very great grief."

The reader will observe that in this case there was a deliberate abandonment of freedom and the company of his own kindred for the sake of human society. There was no coercion. If Jimmie had wished to escape, there was nothing to prevent him, and nothing bound him to his master but an "ever-lengthening chain" of love and aspirations which none but a human being could satisfy.

Here is an instance where a sparrow, one of the most independent and self-reliant of birds, abandoned his own kind for the sake of human beings:

"A lady, whom we know, tells rather a strange story about a sparrow.

"Her brothers had rescued the bird from some boys who had been robbing the nest. They brought it home, and it was reared in the house. It was never confined in a cage, but was allowed to fly freely about the house. As a cat was kept, she had to be watched lest she should injure the bird.

"On Sundays, when the whole family went to church, and no one was left to keep an eye on the cat, the sparrow was always turned into the garden, where he flew about until the family returned. The signal for his entry into the house was that his mistress opened the dining-room window, and stood there *without her gloves*. If she wore her gloves, the bird refused to enter."

A somewhat similar instance is here given, the narrator being an artisan:

"Forty years ago I was in Scotland, living with an uncle at an old castle called Cakemuir. There was a part in ruins, tenanted by quantities of pigeons, many of which were taken for pies. Among them was an unfledged young one, and I, then a boy, took compassion on the solitary thing, and begged it as a pet. I put it in a basket in an empty room, and fed it by hand; and it grew apace, and formed an everlasting friendship for me. It was a bright blue bird, with white head and wings.

"When it was fledged, I gave it liberty, but it would never associate with its fellows. It followed me wherever I went, even for miles, taking long flights, and returning to settle on my arms, head, or shoulders. It was a constant attendant in the breakfast-parlor, driving out the dogs and cats by blows of its wing.

"We removed to another house, where it was perfectly at home. There also it was a great pet with my uncle and aunt, but it would never follow them. After a time I was apprenticed at a village a few miles off, and used to return on Sunday morning and spend the day there. 'Turey' followed me as usual on the Sundays, and when I returned on Monday would try to accompany me. At first I had to drive it back by throwing stones toward it; but it soon learned my intentions, and would only go with me as far as the road. On being told to go home, it would fly around my head, then make a great round in the air and fly home.

"Unfortunately it became troublesome, as most pets do, and used to get into the dairy and disturb the milk. My aunt shut it up, but forgot to give it any water, and the poor bird died of thirst. Many tears were shed, and we were obliged to let my uncle think that I had the bird with me in the village."

I am acquainted with two jackdaws, which behave in much the same manner. One of them entirely declines all intercourse with the jackdaw world, and attaches himself exclusively to the inhabitants of the house. He has the full use of his wings, but generally employs them in flying about the house, and occasionally settling on the heads of persons to whom he chooses to take a fancy. I have had him on my head many times, and it was sometimes rather startling, when absorbed in a book or conversation, to see something black dash before one's eyes, to hear a loud squall of "Jack!" in one's ears, and then to feel the grasp of sharp claws on the top of the head.

The other jackdaw owns a divided allegiance. He does not enter the house, and freely consorts with his fellows. But he is always within, or, at all events, in sight of the garden, and is ready to greet any members of the family who leave the house. He will generally accompany them in their walks; and if they are accompanied by friends who are not acquainted with his ways, he is apt to startle them by an occasional swoop close to their heads, accompanied by a loud caw.

Here is a case of divided companionship in a rook. The anecdote was communicated to me by a lady:

"In the early part of 1861 a young rook was brought to one of my children. It was wounded in the wing, and unable to fly; but every care was taken of it, and it soon recovered. We gave the bird its liberty; but during the whole of that year it kept about the garden and close to the house, always coming to be fed when called.

"He remained with us for some years, when he suddenly disappeared. We feared that he might have been shot; but, to our surprise, about the month of June 'Jack' again made his appearance, sitting in his accustomed place in a tree opposite the window. From that time he has been a constant attendant, coming to us when we call him, and following us from place to place. At other times he joins his companions, and flies about with them, only returning to us to be fed."

The following pathetic little tale shows how the love of master in a dog survived death. It shows reasoning and self-denial on the part of the dog, and also affords another example of the manner in which the power of reason in an animal seems to break down just where it might be expected to manifest itself most successfully:

"Some years ago a fearful snow-storm happened in the Isle of Skye. A shepherd had occasion to go to look after his flock, attended by a faithful dog. The storm increased, and the poor shepherd could not accomplish his task; night had set in, and he was unable to return to his home. Struggling in vain through the drift and darkness, he became utterly exhausted, lay down and died.

"The dog, more fortunate than his master, got back to the lonely sheiling; and when it was seen that he was alone, search was made, but in vain. Hope was giving way to despair, when it was observed that the dog daily took away a piece of bannock, or cake, in his mouth, as it was thought to hide it for some future occasion. But, with that noble instinct with which a wise Providence had endued him (although in this instance unavailing), he set off day by day with this supply to where his master lay, and on being followed it was found that he had placed no fewer than five pieces of bread on his breast. Alas! the vital spark had long since fled, but there was the striking token of instinct and affection, meet subject for even a Landseer to depict."

The intensity of the love which the lower animals can entertain toward man may be estimated from the fact that they have been known to die for the loss of those whom they love. I give three instances of such potent grief, two being exhibited by dogs, and the other by a canary which lived in my own family for some years. The first anecdote is taken from the well-known "Memorials of a Quiet Life," by Augustus Hare:

"Her poor old dog, 'Phloss,' pined away from the moment of his mistress's death. He pined and vexed himself whenever the undertakers came to the house, and on the night before her funeral laid himself down and died—died, as the servant said, just like his mistress, with one long gasp of breath. Thus ended a life bound up in our recollections with 'Julius,' with Havelock, from whom it derived its name, and Julius's dear friend, Tom Starr, by whom it was given."

Then there was my dear dog "Rory," the quaintest, funniest, and most eccentric dog that I ever knew. A rough Irish terrier, black as night, with a triangular patch of snowy white on his breast, and another on the under side of his tail-tuft; thick, heavy eyebrows, with a bold curve in them, only letting the gleam of the glittering eyes sparkle from between their fringes; black moustaches to match the eyebrows, only very much longer and thicker; and ears standing nearly upright for half their length, and then abruptly drooping as if made of black velvet.

I call him *my* dog, not because he ever belonged to me, but because he was pleased to adopt me as his master, and totally to repudiate his legal owner, who, by the way, very honorably paid the tax for him.

Shortly after taking my degree, I accepted a scholastic offer which took me into Wiltshire, where it was impossible to introduce Rory. So, with many regrets, I left him to the care of the household, all of whom were very fond of him.

Of course, he was greatly troubled at my absence, and was perpetually on the watch for me, but after some weeks he seemed to understand the state of things and to be reconciled to his lot. It so happened that after I had been away for some three months, I had to attend to some family business, and visited home for a few hours. Rory was there, and gave me the most curious welcome imaginable.

Naturally a dog of the most exuberant spirits, exalted to the skies by a kind word, and crawling on the ground in utter abasement if scolded, he might have been expected to be more than usually demonstrative when I unexpectedly made my appearance. But he did nothing of the kind. He licked my hand, and that was all. But he would not lose sight of me. He followed me silently about the house, and, when I sat down, lay on the floor, with his chin resting on my foot, and his beautiful loving eyes gazing steadily and wistfully at me through their heavy fringes. He seemed to know that it was for the last time, and kept his steady gaze until I was obliged to leave the house. He made no particular demonstration when I bade him farewell; but his lawful owner claimed him, took him away, and in a few weeks my poor Rory was dead.

There are several now living who will always cherish an affectionate regard for Rory and his odd ways. No human being could have possessed a keener sense of humor than had Rory, and no one could have been more fertile in hitting upon plans for gratifying that sense of humor. He would knock over every fat lap-dog that he met, frighten their mistresses half out of their senses, walk by their sides on his hind legs the whole length of a street, and altogether comport himself like an amiable maniac. He chiefly exulted, however, in alarming college dons as they statelily sailed along in the full glories of silken gown, cassock, and scarf. Such, at least, was the custom in my time, now some thirty years ago; but I am given to understand that in these degenerate days the undergraduates wear moustaches, and a don looks like any body else.

Perhaps that very sense of the ridiculous which was gratified by seeing so stately a being lose all his dignity in instant and groundless alarm was owing to the susceptibility of disposition which, on the one side, hurried him into absurd extravagances, and, on the other side, cost him his very life in disappointed longings for the presence of his self-chosen friend.

The case of the canary was as follows:

It belonged to the head nurse, and was kept in the day-nursery with the children. At all meal-times the cage was always placed on the table, and the bird received much notice. It so happened that the children went away for a few weeks' visit. Although the nurse had the bird in her room, it pined for the society of the children, refused to eat, and in a day or two was found dead at the bottom of the cage.

The following story is related by Mrs. Webber in "The Song-birds of America," and shows how a bird actually died because he thought that he had lost the love of his mistress.

Mrs. Webber had just lost a pet thrush, and was inconsolable. However, a piping bullfinch was brought as a present, and liked to teach her the airs which he knew. At first the bereaved lady would not listen to him, but his winning ways quite overcame her.

"Although I still said I did not love him, yet I talked a great deal to the bird; and as the little fellow grew more and more cheerful, and sang louder and oftener each day, and was getting so handsome, I found plenty of reasons for increasing my attentions to him; and then, above all things, he seemed to need my presence quite as much as sunshine; for if I went away, if only to my breakfast, he would utter the most piteous and incessant cries until I returned to him; when, in a breath, his tones were changed, and he sang his most enchanting airs.

"He made himself most fascinating by his polite adoration; he never considered himself sufficiently well dressed; he was most devoted in his efforts to enchain me by his melodies; art and nature both were called to his aid, until, finally, I could no longer refrain from expressing in no measured terms my admiration. He was then satisfied, not to cease his attentions, but, to take a step further, he presented me with a straw, and even with increased appearance of adulation.

"From that time he claimed me wholly; no one else could approach the cage; he would fight most desperately if any one dared, and if they laid a finger on me his fury was unbounded; he would dash himself against the bars of his cage, and bite the wires, as if he *would* obtain his liberty at all hazards, and thus be enabled to punish the offender.

"If I went away now, he would first mourn, then endeavor to win me back by sweet songs. In the morning I was awakened by his cries, and if I but moved my hand his moans were changed into glad greetings. If I sat too quietly at my drawing, he would become weary, seemingly, and call me to him; if I would not come, he would say, in gentle tone, 'Come-e-here! come-e-here!' so distinctly that all my friends recognized the meaning of the accents at once; and then he would sing to me.

"All the day he would watch me: if I were cheerful, he sang and was so gay; if I were sad, he would sit by the hour watching every movement; and if I arose from my seat, I was called 'Come-e-here;' and whenever he could manage it, if the wind blew my hair within his cage, he would cut it off, calling me to help him, as if he thought I had no right to wear any thing else than feathers; and if I *would* have hair, it was only suitable for nest-building. If I let him fly about the room with the painted finch, he would follow so close in my footsteps that I was in constant terror that he would be stepped upon, or be lost, in following me from the room.

"At last he came to the conclusion that I could never build a nest. I never seemed to understand what to do with the nice materials he gave me; and when I offered to return them, he threw his body to one side, and looked at me so drolly from one eye that I was quite abashed. From that time he seemed to think I *must* be a *very* young creature, and most assiduously fed me at stated periods during the day, throwing up from his own stomach the half-digested food for my benefit, precisely in the manner of feeding young birds.

"But I did not like this sort of relationship very much, and determined to break it down; and forthwith commenced by coldly refusing to be fed, and, as fast as I could bring my hard heart to do it, breaking down all the gentle bonds between us.

"The result was sad enough. The poor fellow could not bear it: he sat in wondering grief—he would not eat; at night I took him in my hand, and held him to my cheek: he nestled closely, and seemed more happy, although his little heart was too full to let him speak. In the morning I scarcely answered his tender love-call, 'Come-e-here;' but I sat down to my drawing, thinking if I could be so cold much longer to such a gentle and uncomplaining creature.

"I presently arose and went to the cage. Oh, my poor, poor bird! he lay struggling on the floor. I took him out—I tried to call him back to life in every way that I knew, but it was useless; I saw he was dying—his little frame was even then growing cold within my warm palm. I uttered the call he knew so well; he threw back his head, with its yet undimmed eye, and tried to answer; the effort was made with his last breath. His eye glazed as I gazed, and his attitude was never changed. His little heart was broken. I can never forgive myself for my cruelty! Oh, to kill so gentle and pure a love as that!"

Many of my readers will anticipate the subject of the next few pages, namely, "Grayfriars' Bobby," a dog whose love of its master long survived death. I have been acquainted with the story of this faithful animal for many years—long, indeed, before the touching narrative was made public through the very prosaic medium of the tax-gatherer.

In the *Scotsman* of April 13, 1867, the following narrative appeared:

"A very singular and interesting occurrence was yesterday brought to light in the Burgh Court by the hearing of a summons in regard to a dog-tax. Eight and a half years ago, it seems a man named Gray, of whom nothing now is known, except that he was poor, and lived in a quiet way in some obscure part of the town, was buried in Old Grayfriars' Church-yard. His grave, leveled by the hand of time and unmarked by any stone, is now scarcely discernible; but, although no human interest would seem to attach to it, the sacred spot has not been wholly disregarded and forgotten. During all these years the dead man's faithful dog has kept constant watch and guard over the grave; and it was this animal for which the collectors sought to recover the tax.

"James Brown, the old curator of the burial-ground, remembers Gray's funeral, and the dog, a Scotch terrier, was, he says, one of the most conspicuous of the mourners. The grave was closed in as usual, and next morning 'Bobby,' as the dog is called, was found lying on the new-made mound. This was an innovation which old James could not permit; for there was an order at the gate stating, in the most intelligible characters, that dogs were not admitted. 'Bobby' was accordingly driven out; but next morning he was there again, and for the second time was discharged. The third morning was cold and wet; and when the old man saw the faithful animal, in spite of all chastisement, still lying shivering on the grave, he took pity on him, and gave him some food.

"This recognition of his devotion gave 'Bobby' the right to make the church-yard his home; and from that time to the present he has never spent a night away from his master's tomb. Often in bad weather attempts have been made to keep him within doors, but by dismal howls he has succeeded in making it known that this interference is not agreeable to him, and latterly he has always been allowed his own way. At almost any time during the day he may be seen in or about the church-yard; and no matter how rough the night may be, nothing can induce him to forsake the hallowed spot, whose identity, despite the irresistible obliteration, he has so faithfully preserved.

"Bobby has many friends, and the tax-gatherers have by no means proved his enemies. A weekly treat of steaks was long allowed him by Sergeant Scott, of the Engineers; but for more than six years he has been regularly fed by Mr. Traill, of the restaurant, 6 Grayfriars Place. He is constant and punctual in his calls, being guided in his mid-day visits by the sound of the time-gun. On the ground of harboring the dog, proceedings were taken against Mr. Traill for payment of the tax. The defendant expressed his willingness, could he claim the dog, to be responsible for the tax; but so long as the dog refused to attach himself to any one, it was impossible to fix the ownership; and the court, seeing the peculiar circumstances of the case, dismissed the summons.

"Bobby has long been an object of curiosity to all who have become acquainted with his history. His constant appearance in the grave-yard has caused many inquiries to be made regarding him, and efforts without number have been made to get possession of him. The old curator, of course, stands up as the next claimant to Mr. Traill, and yesterday offered to pay the tax himself rather than have Bobby—Grayfriars' Bobby, to allow him his full name—put out of the way."

Four years longer the faithful little dog kept his loving watch, and at last died, to the regret of all who knew him, never having been out of reach of his master's grave; though in his later years the infirmities of doggish age forced him to accept a partial hospitality of the curator. I am sure that Lady Burdett-Coutts gladdened the hearts of many lovers of animals—as she certainly did mine—when she perpetuated his memory by a lasting monument of granite and bronze. The monument is a drinking-fountain made of Peterhead granite, and surmounted by a life-size statue of Bobby in bronze.

During the many years which elapsed between the death of his master and his own departure, the lowly grave was forgotten by all but the dog. No stone guarded it, and not even a mound marked it. The grass and weeds grew luxuriantly over it as over the level soil around. There has been for years nothing that could mark out the grave from the surrounding soil, but the little dog knew the sacred spot under which lay his master's remains, and for hours used to stand upon it, keeping his guard. A little way from the grave is an altar-tomb, under which Bobby used to shelter himself in bad weather, and to which he always used to take the bones and other food provided for him by the generous persons whose names have already been mentioned.

I possess three photographic portraits of Bobby. One represents him as standing upon the nameless grave, which is utterly indistinguishable from the weeds and herbage around. The portrait is not quite so good as it might be; for just as the photographer had got the dog into focus, and had uncovered the lens, Bobby unfortunately caught sight of a dog passing the gate of the church-yard, and, according to custom, flew at him furiously. He did not seem to object to human beings, but a dog he never would permit to be even in sight.

The best of the three portraits is that from which the bronze statue has been taken. He is sitting on the altar-tomb above mentioned, and is looking upward with that wistful, patient, longing, yearning expression of countenance which was peculiar to the animal, and is conspicuous in all the photographs, however imperfect they may be.

Some animals, notably dogs, have a wonderful power of returning to their beloved master, even though they have been conveyed to considerable distances. So many examples of such feats are on record that I refrain from mentioning them, and only give one or two, the truth of which is guaranteed by my correspondents, whose letters I possess.

"A gentleman in Calcutta wrote to a friend living near Inverkeithing, on the shores of the Firth of Forth, requesting him to send a good Scotch collie dog. This was done in due course, and the arrival of the dog was duly acknowledged. But the next mail brought accounts of the dog having disappeared, and that nothing could be seen or heard of him. Imagine the astonishment of the gentleman in Inverkeithing when, a few weeks later, friend Collie bounced into his house, wagging his tail, barking furiously, and exhibiting, as only a dog can, his great joy at finding his master.

"Of course all inquiry was made to find out how Collie got home again, when it was discovered that he had landed from a collier which had returned from *Dundee*. Inquiry was made at Dundee, when it was found that the dog had come there on board a ship from Calcutta. Now it can be understood that the dog might have recognized the collier, as he might have seen the vessel on some former occasion at Inverkeithing; but how he should have selected, at Calcutta, a ship bound for Dundee is not so easily explained."

There is one solution of this remarkable problem which has occurred to me. Probably the dog, not liking the strange land and the dark faces, had slipped back to the ships with which he had been familiar at home. Recognizing the well-known Scotch accent on board one of the ships, he must have got quietly on board, and, on landing at Dundee, transferred himself to the collier. This is merely conjecture, but I do not see any other mode of accounting for the dog's wonderful journey.

A scarcely less wonderful feat was performed some time ago by a dog which returned to his mistress from a distance. It is true that Manchester is not so far from Holywell as India is from Scotland; but the journey, though shorter, was very much more complicated, and involved several modes of locomotion, some of which, at least, must have been adopted by the dog. The narrator of the story is my friend, the late J. Hatton, M.D., whose name has been perpetuated on a life-boat presented by his widow to the Dungeness station:

"Some years ago, when I lived in Manchester, I attended, for fever, a mechanic, who worked for Messrs. Sharp, Roors, & Co., the celebrated locomotive-engine makers. When he became convalescent, he went to the house of his mother, who then lived at Holywell, in Wales. After he had recruited his health, and was about to return home, his mother gave him a dog.

"He led the animal from Holywell to Bagill by road, a distance of about two miles. Thence he took the market-boat to Chester, a distance of about twelve miles, if I remember right. Then he walked through Chester, and took rail for Birkenhead. From that station he walked to the landing-stage and crossed the Mersey to Liverpool. He then walked through Liverpool to the station at Lime Street. Then he took rail to Manchester, and then had to walk a distance of a mile and a half to his home.

"This was on Wednesday. He tied the dog up, went to his work on Thursday as usual; and on the Sunday following, thinking that the dog was accustomed to the place, he set it at liberty. He soon lost sight of it, and on the Wednesday following he received a letter from his mother, stating that the dog had returned to her. Now you will see that the dog went first by road, then by market-boat, then through streets, then by rail, then by steamer, then through streets again, then by rail again, and then through streets again, it being dark at the time; and yet the dog had sagacity enough to find its way back to the scene of its early recollections."

In this, as in other stories of a similar character, one of the most curious points is the extreme rapidity with which the animal made the journey. I do not know whether the market-boat ran on Sunday, but, at all events, the dog must have achieved the distance in some forty-eight hours.

That the dog in question retained a remembrance of the route by which he had traveled, and knew how to avail himself of the means of transit, I have no doubt whatever; and this notion is confirmed by the behavior of a dog that belonged to one of my correspondents, Mr. B., who has kindly sent me several dog-biographies that came within his own experience.

He was then living in East Lothian, and had given the dog, a prize greyhound, to a friend who lived at Greigston, near Cupar, in Fife. His new owner took him home, but in a few days the dog was missing. His owner advertised his loss, and the animal was captured on the pier at Burntisland, evidently waiting for an opportunity to cross in the steamer, whence he would undoubtedly have found his way back. This pier is fully twenty miles from the spot which he deserted.

I can easily understand how a dog would manage to slip on board by pretending to belong to one of the passengers. Dogs are quite alive to the social distinction between those who belong to some particular owner and those who are masterless, the latter being looked upon by themselves much as a "masterless man" was regarded in the time of Elizabeth, *i. e.*, a sort of social outcast, unacknowledged by his fellows.

I owed the life-long friendship of my inimitable Rory to this feeling. He was none of my dog. He belonged to a man of another college, with whom I had hardly exchanged half-a-dozen sentences. His master was obliged to cross the sea during the long vacation, and left the dog in the charge of his scout.

Being always of an aristocratic turn of mind, Rory repudiated the scout altogether, and, remembering that he had been in my rooms at Merton, he paid me a visit one morning, and engaged me as his master. It was not the least use to take him back, for he always returned in an hour or two; and at last it was tacitly agreed that he should retain possession of me. He knew the value of a collegiate master, and was not going to be fobbed off with a scout. His legitimate master having deserted him, he exercised his right of selecting a master for himself, and accordingly he chose me, and kept me, and when we were parted he died of grief, as has already been narrated.

I know another dog who displayed great wisdom in escaping the snares of London life.

He was a beautiful little dog, just the animal whom a professional dog-stealer would be sure to snap up, if possible. One day he had been for a drive with his mistress, and, on being allowed to alight with her, had in some way been separated from her. After a vain search she drove home, and sent the servants to try and find her lost pet. He was presently discovered by the coachman trotting quietly homeward, not in haste like a lost dog, but with a composed air, and pretending that he belonged to some one who was going in the same direction.

I have seldom met with a more curious example of the ability of a dog to find his master than is related in the following story:

"Some years ago, when I lived in Fife, I was coming to Edinburgh with my eldest daughter. Zeno accompanied us to the station, about a mile from home; and as I did not wish him to come any farther, I asked a gentleman who was living with me to take him home.

"Just as the train was about to start I looked out to see if he had gone, when I saw him following my friend up the stairs at the station. We rattled along for a distance of about twelve or fourteen miles till we reached Burntisland, where we had to cross the Forth.

"The day became very cold, with snow and sleet, so we hurried down to the steamer. We had left the pier about ten minutes when a passenger, wishing shelter, opened the saloon-door, when, to my great surprise, in walked Zeno, sniffing his way up till he came to me and jumped upon my knee. How he came was a mystery to me, and ever will be. All the carriages were shut when I last saw him. I think that he must have returned and got into the guard's van; but no one could tell me, and the strange thing was that he did not get out at any intermediate station.

"I still have the old dog, and he is as dear to me as ever. Never was there his like: never did he bite, though teased by children and grandchildren. His life is now one of constant repose; and when the cord will one day snap which will sever our long and faithful connection, I shall mourn his loss as that of a friend.

"Talk of 'dumb animals'—we might well take lessons from them in many things; they would even put many to shame. Yes, ours is a higher and a nobler destiny; but yet, withal, methinks we might learn to profit from much we both know and hear of in the lives of our animal friends and relations."

The reader will perhaps remember that Zeno has already been mentioned under the head of "Jealousy."

It is often, but erroneously, said that cats are selfish animals, attaching themselves to localities and not to individuals. This idea has, perhaps, some ground of truth, for it is not so easy to understand the nature of a cat as that of a dog; and when a cat is not understood, it is very likely that she cares less for the inhabitants of a house than for the house itself. But I know of many instances where cats have been in the habit of moving about with their owners, and have been as unconcerned as dogs would have been.

My own cat "Pret," for example, was first taken from a small house at Greenwich to a large one in the very heart of the city, where he had the range of many cellars, but no open air. Next he went to another large house in the city, where he had no cellars, and could only get on the leads by special permission. Then he was moved into a house in the country, where he had neither cellars, leads, nor tiles, but a garden. After that we moved to a larger house in the same village, whither he followed us of his own accord.

His mother, "Minnie," always accompanied her mistress when she was on a visit, and I have more than once taken Minnie to her mistress for a journey of several miles. Here is a corroborative letter from a lady:

"I believe, for my part, that cats attach themselves to people and not to places. Our cats always seemed to know their masters. One, belonging to my sister, would scratch all the rest of the family, though quite gentle with her. We traveled about once for a year and a half with a favorite cat; though during that time we changed our lodgings many times, she never left us. She even seemed to know our rooms, and kept to them when there were other apartments in the same house. She used to knock at the door when she wanted to come in, and would endeavor to turn the handle by taking it between her paws. I have also seen her, when she was thirsty and could not reach the water in the jug, dipping her paws in to get it in that way. She would follow my brothers around the room when they whistled a tune, and rub her head against their hands and face, and touch their lips with her paw, as if pleased with the sound."

Perhaps the reader may remember the history of Lady E.'s cat Rosy, on p. 109, in which it is incidentally mentioned that she always traveled with her mistress.

My late esteemed friend, Mr. W. Chambers, called my attention some years ago to a story of a cat, which showed that the attachment of the animal toward man is much stronger than toward locality. He guaranteed the truth of the statement, and furnished me with the name and address of the person to whom the cat belonged.

The story is briefly as follows:

A man and his wife, living in the northern part of Scotland, near the west coast, had to move to a place on the east coast. In consequence of the expense of taking furniture by land, they traveled by sea, passing around the northern point, and landing near their new home. Having been told that cats only cared for localities and not for human beings, they, meaning all kindness, left the animal behind them. They, however, had not been long settled in their new home, when the man, on returning from work, saw a cat sitting on the wall, and found that it was actually his own cat, who, by some mysterious means, had found him out. She was hungry, emaciated, and tired, and had evidently traveled by land to the same spot which they reached by sea. The power by which she did it may be instinct, or it may be the exercise of a faculty not possessed by man. But I have related the anecdote to show how great must have been the love felt by the cat for its master, when it left the home which it knew well, and took a long and fatiguing journey to join its master in a house which it did not know.

The following anecdote shows that the cat does love people more than places:

"Last summer we were staying for some weeks at Victoria Place, Eastbourne, and every morning the town-crier came in front of our house, giving out the public amusements for that evening, and a list of articles lost. Judging from the large number of things daily missing, either visitors or inhabitants must have been a most careless race. He was the most amusing crier I ever heard, making his announcements in a semi-dramatic style and tone, which, together with a good voice and most pompous delivery, rendered these minor affairs quite important. One of the missing items especially attracted my attention.

"'Lost, a tortoise-shell cat, of the Persian breed, with a velvet collar around its neck, rather old and very shy. Whoever will bring it to the Crier, dead or alive, shall receive ten shillings reward.'

"This was repeated for many days, and then the reward was increased to one sovereign, with the intimation that no larger sum would be offered. At about the end of three weeks the 'Cat' was taken off the list, and I inquired after the fate of poor pussy.

"The cat, which was of rare beauty, had been brought to this country as a present to Lady ——, and had for years accompanied her when traveling. Soon after their arrival at Eastbourne, that love of liberty inherent in all animals, and a due appreciation of the surrounding scenery, induced pussy to stray into the woods, where she was at first hunted as a wild cat, and afterward chased and shot at to obtain the offered reward. She contrived to escape all these dangers, and existed on the few wild birds that she could catch, until Lady —— heard of her whereabouts and went in search of her. The poor half-starved pet, on hearing the voice of her mistress calling her name, jumped on her shoulder, and thus terminated her rambles in the wild woods. It is a most dangerous mistake to offer a reward for a lost pet, 'dead or alive'—the addition of the former word, while facilitating its capture, oftentimes proving its death-warrant."

The same lady, who communicated the preceding anecdote, has favored me with two more, showing the attachment felt by dogs to their masters:

"A friend of ours, a great traveler, who has generally several dogs of various breeds, always takes one of them with him, making it a rule to take a different dog each voyage, in order, as he says, 'to give them all a foreign education;' home occupations preventing him from bestowing much time upon them when in England. Our dogs do not have this advantage, being generally left at home during our absence in charge and under the tuition of an old housekeeper.

"A remarkable instance of the power of scent was manifested by our little Maltese dog, 'Joey.' Our travels are often long in duration, and far distant; but, however numerous the post-offices through which our letters passed, he could always distinguish them from others, evincing great joy when allowed to smell them, and often trying to obtain possession. He was anxiously watching the postman's knock one morning, when several letters arrived. Accidentally they all fell to the ground. Joey took advantage of his position, selected ours, and rushed off in great glee, giving the old housekeeper a famous run around the garden, and then most decidedly refusing to give up his prize. She was obliged to adopt the expedient of slipping the letter (slightly damaged) out of the envelope, and allowing him to retain the latter, which he carried off in triumph to his basket."

This is the same dog of whom several anecdotes have already been related, showing his great mental capacities.

As far as I can learn, all animals have not only a capacity for the society of men, but an absolute yearning for it. This feeling may be in abeyance, as having received no development at the hands of man; but it is still latent, and may be educed by those who are capable of appreciating the character of the animal. Tigers, for example, are not generally considered the friends of mankind, and yet the Indian fakirs will travel about the country with tame tigers, which they simply lead with a slight string, and which will allow themselves to be caressed by the hands of children without evincing the least disposition to make a meal of them.

In the case of domestic animals, even the fiercest of them have this innate longing for human society, and will indulge it when they have the chance. This chance generally occurs by means of confidence on behalf of the human being. The animal is surprised to see some one who is not afraid of him, and so gives his confidence in return. The reader will perhaps remember how that eminently ferocious "Mess" became my very good friend. A somewhat similar case has just been related to me.

A little girl, about two years old, wandered from her nurse, and was lost. At last the child was found asleep in the kennel of a peculiarly savage bloodhound, named "Rob." The dog was jealously guarding his little charge, and would let no one approach until the mother came. She called the child, who came to her, followed by Rob. She took the child home, but Rob insisted on accompanying them; and as they went, the little girl held her mother with one hand and Rob's ear with the other—the child being so small, and Rob so large, that he had to walk all the way with his head bent down.

I have read an account, but do not recollect where, of a boy who went into a stable inhabited by a notorious "savage." He did not know the character of the animal, gave him bread and other delicacies, and the horror of the groom may be imagined when one day he found the boy and the horse lying together on the floor of the stable—the boy not having the slightest idea of the character of the horse, and the horse not having the least intention of hurting the boy, but cherishing him as a valued companion.

I have now the pleasure of giving a few little histories showing the affection which is often entertained for man by animals which he is not generally accustomed to consider as his companions. I have already mentioned an instance of friendship between a sheep and cows, and I now give two examples of the same attachment of sheep to man:

"We had a pet lamb, which was fed by the cook. When the lamb was about six weeks old, the cook became ill, and was confined to bed for some days. While she was ill, the lamb left its usual place of abode, lay beneath her bed, and refused all food, although the milk was offered from the usual bottle. It did not seek nor worry the sick servant, but lay perfectly quiet under her bed.

"A pet sheep of my late brother has come to end its days with us at Bassendean. This sheep was the constant out-door companion of my brother and his niece. They were, however, obliged to give up walking with him, for he would insist on pushing his way between them, and would not condescend to walk on one side."

We are rather apt to consider the goose (including gander) as a peculiarly stupid bird, and to use its name, as we do that of the ass, as a synonym for folly. Yet a greater mistake could not be made in either case. We have already been told of an ass which his master was obliged to sell because he was too clever to be kept, and led the other animals into mischief. The same writer now tells us a story of affection between a goose and a man:

"A goose—not a gander—in the farm-yard of a gentleman, was observed to take a particular liking to her owner. This attachment was so uncommon and so marked that all about the house and in the neighborhood took notice of it; and, consequently, the people, with the propensity they have to give nicknames, and with the sinister motive, perhaps, of expressing their sense of the weak understanding of the man, called him 'Goosey.' Alas for his admirer, the goose's true love did not yet run smooth; for her master, hearing of the ridicule cast upon him, to abate her fondness, insisted on her being locked up in the poultry-yard.

"Well, shortly after he goes to the adjoining town to attend petty sessions; and in the middle of his business what does he feel but something wonderfully warm and soft rubbing against his leg; and on looking down he saw his goose, with neck protruded, while quivering her wings in the fullness of enjoyment, looking up to him with unutterable fondness. This was too much for his patience or the by-standers' good-manners; for, while it set them wild with laughter, it urged him to a deed he should ever be ashamed of; for, twisting his thong-whip about the goose's neck, he swung her round and round until he supposed her dead, and then he cast her on the adjoining dunghill.

"Not very long after, Mr. Goosey was seized with a very severe illness, which brought him to the verge of the grave; and one day, when slowly recovering, and allowed to recline in the window, the first thing he saw was his goose, sitting on the grass, and looking with intense anxiety at him. The effect on him was most alarming.

"'What!' says he, 'is this cursed bird come back to life? and am I, for my sins, to be haunted in this way?'

"'Oh, father,' says his daughter, 'don't speak so hardly of the poor bird. Ever since your illness it has sat there opposite your window; it scarcely takes any food.'

"Passion, prejudice, the fear of ridicule, all gave way before a sense of gratitude for this unutterable attachment. The poor bird was immediately taken notice of, and treated from henceforth with great kindness; and, for all that I know, goose and Goosey are still bound in as close ties as man and bird can be."

The second story, told by the same writer from his own observation, has such a tragic conclusion that I could hardly make up my mind to print it. There is in both cases extraordinary love, amounting indeed to worship, on the part of the bird toward man. In both cases there is not only a want of reciprocity, but actual ingratitude, on the part of the man toward the bird—redeemed, however, in the preceding anecdote by repentance and reciprocation of friendship:

"I must tell you, among many anecdotes I know of geese, one that came under my own observation. When a curate in the county of Kildare, my next neighbor was a worthy man who carried on the cotton-printing business, and who, though once in very prosperous circumstances, was now, in consequence of a change in the times, very poor.

"In his mill-yard was a gander, who had been there forty years; he was the largest bird of his kind I ever saw. His watchfulness was excessive: no dog could equal him in vigilance, neither could any dog be more fierce in attacking strangers or beggars; he followed his old master wherever he went, and at his command would fly at any man or beast; and with his bill, wings, and feet he could and would hurt severely.

"Whenever my neighbor paid me a visit, the gander always accompanied him; and as I was liberal of oats, and had besides one or two geese in my yard, he would, before his master rose in the morning, come up and give me a call; but neither the oats nor the blandishments of the feathered fair could keep him long away, and he soon solemnly stalked back to his proper station at the mill.

"Well, year after year I was perfecting my friendship with Toby, the gander, and had certainly a large share in his esteem; when one winter, after being confined to the house with a severe cold, I, in passing through the mill-yard, inquired for my friend, whom I could nowhere see.

"'Oh, sir,' said the man—and he was about the place as long as Toby himself—'Toby's gone.'

"'Gone where?'

"'Oh, he is dead.'

"'How! dead?'

"'Why, we ate him for our Christmas dinner.'

"'Ate him!'

"I think I have been seldom in the course of my life more astonished and shocked. Positively I would have given them a fat cow to eat, could I have saved poor Toby; but so it was. Upon inquiry, I found out that the poor gentleman had not means to buy his Christmas dinner; that he was too proud to go in debt, and, determined as he was to give his people a meat dinner, poor Toby fell a sacrifice to proud poverty. While honoring the man for his independence, I confess I never could look upon him afterward without a sense of dislike. I did not either expect or desire that he should suffer as he who slew the albatross, but I was sure that he would not be the better in this world or the next for killing the gander."

In which sentiment I, and I hope all my readers, most cordially agree.

Surveying all these examples of love displayed toward human beings by animals, it is impossible really to believe that such love can die. Unselfish love such as this, which survives even ingratitude and ill-treatment, belongs to the spirit and not to the body, and all beings which are capable of feeling such love must possess immortal spirits. All may not have an opportunity of manifesting it, but all possess the capacity, and would manifest it openly if the conditions were favorable.

We will just run over the anecdotes which I have given. In those of the squirrel, the sparrow, the blue pigeon, the jackdaws, and the rook, we have examples to show that even in the wild animals the love of human beings can overpower that of liberty and of their own kind, and that they will forsake both liberty and their kinsfolk for the society of man. I have no doubt that this is due to their appreciation of a nature higher than their own, and the feeling that their own nature is purified and elevated by contact with man. Indeed, it is a fact that, whenever man and beast are brought into contact, those which possess natures capable of elevation and development cleave to him, court him, and thrive by his presence; whereas those which are incapable of improvement perish before his presence.

It is the same with the human race. When civilized man comes in contact with a barbarian, the latter rapidly tends toward civilization, throws off his barbarian customs, adopts those of the superior being, learns by degrees his arts and sciences, and so gradually merges into civilization. With the savage the case is different. He can not learn any thing good from the higher race. He may, and does, gain means by which to develop more completely his evil tendencies, but is utterly incapable of improvement. He can neither replenish the earth nor subdue it, and so he perishes before the presence of those who do, at all events, endeavor to carry out that which is the great mission of man. Wherever civilized man sets his foot, the savage dies out.

Why this is we can not say; but it is a fact long familiar to anthropologists. The Tasmanians have all gone. I have portraits of the last three survivors, all of whom have since died. But the strange thing is that the race has died out for want of new births, not because it was extirpated by slaughter. For years before that final extinction of the Tasmanians, no children were born.

A similar phenomenon, though slower in its operation, is now to be seen in New Zealand. The native race, splendid specimens of the savage as they are, become yearly fewer and fewer in the presence of the European, the births falling far short of the deaths. Even in the vegetable world the same idea is carried out, and the grand tree-ferns, as large as our oaks, are perishing before the advance of the English clover. The lower creation, if it can not be elevated by the presence of the higher, dies out, and the same rule holds good with man, with beast, and with plant.

The next division of the subject shows how intense must be the love of animals to man, when the deprivation of the object of their affection has killed them. We sometimes hear of human beings dying from a similar cause, and none of us who heard that a man or woman had died from grief at the loss of a friend would think for a moment that such intensity of love could proceed from any other source than the spirit.

Lastly, we have cases where animals, not usually made the companions of man, have, unsought, conceived a deep affection for human beings, and have cherished that love in spite of neglect, indifference, dislike, and violence. Such a love is utterly unselfish, and must issue from the same source that causes man to abandon the love of self for the love of others. It is, in fact, loving the neighbor better than one's self.

CHAPTER XV.

CONJUGAL LOVE.

Necessary Limits of Conjugal Love among Animals.—Non-pairing Animals.—Polygamous Animals.—Animals which Pair for a Season.—Animals which Pair for Life.—Supply of Spare Partners.—The Turtle-dove, the Eagle, and the Raven.—Conjugal Love in the Teal.—Picture of the "Widow."—Conjugal Love among Fishes.—The "Devil-fish" and its Fate.—The Chocollito of South America.—Faithlessness, Sorrow, and Death.—Materials for Drama.

As may easily be imagined, there are but few animals in which this kind of love can be manifested. The greater number of species have no particular mates, but simply meet almost by chance, and never trouble themselves about each other again. No real conjugal love, therefore, can exist, and it is rather curious that in such animals a firm friendship is often formed between two individuals of the same sex.

Next we come to polygamous animals, such as the stag among mammals and the domestic poultry among birds. Here is a decided advance toward conjugal love, although, as in the case of polygamous man, that love must necessarily be of a very inferior character. Here is, at all events, a sense of appropriation on either side, and, as has been already mentioned in the chapter headed "Jealousy," the proprietor of the hárem resents any attempt on the part of another male to infringe on his privileges.

Next we come to those examples where, as in many birds, a couple are mated for the nesting-season, but do not afterward seem to care more for each other than for their broods of children. If, during the nesting-time, one of the pair be killed, the survivor, after brief lamentation, consoles itself in a few hours with another partner. There really seems to be a supply of spare partners of both sexes always at hand; for, whether the slain bird be male or female, one of the same sex is sure to be ready to take its place.

Lastly, we come to those creatures which are mated for life, and among them we often find as sincere conjugal love as among monogamous mankind. Prominent among them are the eagle, the raven, and the dove; and it is remarkable that while we praise for its conjugal fidelity the turtle-dove, the type of all that is sweet, good, and gentle, we entirely forget to accredit with the same virtue the eagle and the raven, types of all that is violent, dark, and cunning. There are many anecdotes in existence of the conjugal love among such birds, but, as they are so well known, I shall not refer to them, and only mention one or two with which we are not so familiar.

I shall give only three instances, all of which show how deeply conjugal affection can be felt by the lower animals, and how completely the love of self is forgotten in the love of the partner. In the first of these instances, life was risked in the face of danger, and only spared by reason of forbearance; in the second, life was risked and lost; and in the third life was lost without the intervention of any external danger.

In Hardwicke's *Science Gossip* for 1870, p. 36, there is an account of the teal, in which the conjugal love displayed by this bird is well shown. The writer had been duck-shooting, and had just shot a mallard, when a couple of teal sprang up, alarmed at the report.

"The duck, being the nearest, received the contents of the remaining barrel, and fell dead upon the soft mud at the very edge of the water.

"While speculating upon our good luck, and putting in two fresh cartridges, the cock teal, which had flown up to the other end of the pool when his mate fell, turned back, and, after flying up and down several times with mournful notes, returned to the spot whence he rose, and pitched upon the mud, close to the dead duck. Here he remained for some seconds, nodding his head and courtesying, as if about to take wing, uttering a low note the while, as if to entice away the duck, whom he appeared so loth to leave.

"We were so struck at this manifestation of affection that we could not find it in our heart to shoot the poor bird; and, as we moved on to pick up his mate, he rose, and was soon out of range again."

Perhaps the reader may remember a beautiful painting by Landseer, entitled "The Widow," in which a similar scene is represented, except that it is the drake which is lying dead, and the duck which is mourning over her deceased partner.

Fishes are thought to be rather prosaic beings. They do not possess much expression of feature, at all events, to human eyes; and their habits and their looks are, as a rule, much on a par. Yet there is at least one instance known in which a fish, and that a singularly hideous one, exhibited a degree of conjugal love which would have done honor to any human being.

Inhabiting the waters of the Mediterranean Sea is a gigantic ray, called popularly the Devil-fish, and scientifically *Cephaloptera Massena*. These fishes are formed much like our common ray, but attain the most enormous dimensions, sometimes measuring thirty feet across the fins. The power of this fish is quite proportioned to its size. When pierced with eight or ten harpoons, and towing behind it a string of as many boats, all pulling against it, the devil-fish has been known to drag the whole line some ten miles to sea, and finally to break lose and escape, with all the weapons still sticking in its back.

The Mediterranean fishermen employ in the capture of the tunny a vast net, called a mandrague, which is separated into many chambers. In one of these nets a female devil-fish contrived to entangle herself, was captured and taken ashore. She weighed 1328 lbs. A male who had accompanied her, but had not got into the net, was disconsolate at her capture, and for two days haunted the spot where his companion had been captured. He wandered round and round the nets, seeking for his lost mate, and was at last found in the mandrague, but dead, having died of grief.

The last case is that of some little South American parrots, called Chocollitos. They are charming little birds, gentle, and easily tamed. They are among the monogamous birds, and are, as a rule, strictly faithful to their marriage vows. There are, however, exceptions to most rules, and one of these is related by Froebel, in his work on South America.

The traveler in question was a guest for a while in a house at Granada. In this house about twenty chocollitos were kept; and, as they were all brought to the house when very young, they did not form their matrimonial attachments until after their arrival. Perhaps among them the sexes were not equally divided, so as to insure each bird a mate; but the sad fact was that, after one pair had entered the marriage state, another male made love to the wife. The lady was weak, and yielded to the solicitations of the too fascinating lover.

The result was, according to Froebel's own words, as follows: "When the husband understood the whole extent of his misfortune, and after he had made the last unsuccessful attempt to bring his faithless companion back to the path of duty, the unhappy creature, heart-broken by his wrongs, took his lonely seat on the perch on which he had passed happier nights, closely pressed to the side of his partner, refused to eat or drink, and one morning was found dead on the floor below."

The reader may compare this narrative with that of the Mandarin duck, narrated on page 94. In both cases there was strong conjugal love; but in the former the lady was faithful, and her husband avenged himself on the disturber of his domestic peace; while in the latter the lady was frail, and the husband died of a broken heart. Both narratives are wonderfully human, and each could furnish the plots of a sensational drama.

CHAPTER XVI.

PARENTAL LOVE.

Absence of Filial Love among Animals.—Analogy with Human Beings.—The Savage and his Parents.—Parental Love among Animals.—The "Storgë" of Theorists.—Identity of the Feeling in Animals and Man.—Endurance of Parental Love in Animals.—Exceptions to the General Rule.—A Cat and Two Generations of Kittens.—My own Cat and her Young.—The Dog "Georgie" and her Daughter.—Abnegation of Self.—The Fly-catcher and my Cat.—A Released Prisoner and Joyous Escape.—A Courageous Swallow.—Redbreast and Viper.—Passive Courage in a Partridge.—The Whale and her Young.—A Duck's Journey, and Rescue of her Young.—Do Animals have Names in their own Language?—The Mystery of Parental Love in Birds.—Love and Intellect.—Parental Love among Fishes.—The Stickleback and its Nest.—Apparent Reversion of Parental Love.—The Pipe-fishes and Sea-horses.—The Cursorial Birds and their Eggs.—A Brave Spider.—Comparison between Man and Animals.

BEFORE beginning this subject, I can not but remark the apparently singular fact that, whereas among the lower animals we find so many instances of the love of parents toward their offspring, we see so few, if, indeed, any trustworthy accounts of the Filial Love, or love of children toward their parents. Yet the same analogy prevails in this as in other cases which have already come before us, and we must look to man if we wish to understand the lower animals. Even human nature must be highly developed before filial love can find any place in the affections. In the savages it barely exists at all, and certainly does not survive into mature years.

Take, for example, even such fine specimens of the savage as the North American Indian and the Fijian. The idea of being subject to their parents never enters their heads; still less does the idea of loving them. It is the glory of a North American Indian boy, at as early an age as possible, to despise his mother and defy his father. And the women are just as bad as the men. They, rejoicing in the pride of youth and strength, utterly despise the elder and feeble women, even though they be their own mothers, and will tear out of their hands the food which they are about to eat, on the plea that old women are of no use, and that the food will be much better employed in nourishing the young and the strong.

Then, if the tribe be on the move, and those who are old and infirm are felt to be hinderances, they settle the matter by leaving them behind. They just salve their consciences by building a slight shelter of sticks and boughs, lighting a fire, and leaving a little water and food. But they know perfectly well that before another sun has set there will be nothing left of their victims but the bones, the wolves having made short work of them as soon as the tribe was out of sight. The forsaken make no complaints, neither do those who press forward expect a better fate; and hence it is that they all wish rather to fall in battle than to die a natural death, after feeling themselves a burden to all around them.

The charming little episode in "Robinson Crusoe," where Friday finds and rejoices over his father, is a very pretty piece of writing, but quite out of accordance with the repulsive reality of savage life.

As to the Fijians, they have not the least scruple in burying a father alive when he begins to be infirm, and assist in strangling a mother so that she may keep him company. With regard to the Bosjesmen of South Africa and the "black fellows" of Australia, I very much doubt whether they ever have possessed the least idea that any duty was owing to a parent from a child. Nor have they much notion of duty from a parent toward the child. The father is just as likely as not to murder his child as soon as it is born—perhaps rather more likely than not; and if he be angry with any one for any reason, he has a way of relieving his feelings by driving his spear through his wife or child, whichever happens to be nearest.

Even the mother treats her child rather worse than a cow treats her calf, and leaves the tiny creature to shift for itself at an age when the children of civilized parents can scarcely be trusted to pass a quarter of an hour alone.

This being the case with parental love, it may be easily imagined that filial affection can have but little opportunity of development, and I very much doubt whether in the true savage it really exists at all in the sense in which we understand it.

As, therefore, we find that in the lower human races filial love either is very trifling, or is absolutely non-existent, we need not wonder that in the lower animals we find but few, if any, indications of its presence.

We now proceed to the subject of this chapter, namely, PARENTAL LOVE, and the various modes in which it develops itself.

There are many writers who assert that parental love in the lower animals is not identical with that of man. They say that it is only a sort of blind instinct, and, in order to mark more strongly the distinction between man and beast, call the parental love of the latter by the name of "storgë." For myself, I really fail to see any distinction between the two, except that in civilized man the parental love is better regulated than among the lower animals. But, as we have already seen, among the uncivilized races it is not regulated at all, and, indeed, many of the beasts are far better parents than most savages.

Neither can I understand why the word "storgë" should be applied to parental love among the lower animals, and not to the same feeling in man. The word is used by Greek writers, together with the verb from which it is formed, to signify the love between human parents and children. For example, in Plato we have the term used for mutual love between parents and children—"The child loves, and is loved by its parents;" and the same word is used in the same sense in several passages of Sophocles and other writers.

One argument which is always employed by those who deny the identity of the feeling in both cases is that parental love endures throughout life in man, while in the lower animals it expires with the adolescence of the young. This statement is partly, but not entirely, true. As a rule, it is true with civilized man; but, as I have already shown, the parental love of a savage does not last longer than that of a bird, a cat, or a dog, taking into consideration the relative duration of life. And the reason is the same in both cases. Were parental love to exist through life in the savage, the bird, or the beast, the race would soon become extinct. Neither is able to support their children longer than their time of helplessness. The beast and the bird can not, and the savage will not, provide for the future; and if the young had to depend upon their parents for subsistence, they would soon perish of hunger.

There are, however, exceptions to this general rule, and always, as far as I can see, in domesticated animals whose means of subsistence are already insured. Several of such cases have lately come before my notice. One has been already narrated under a different heading, i. e., "Sympathy," p. 110, where some traits of two cats, a mother and daughter, are recorded. I here present the reader with another anecdote of parental love surviving adolescence. It is a very remarkable story, because we see, in the first place, the usual law prevailing, and the once-favorite child driven away in anticipation of a new family. That family having perished, the original parental love resumed its sway, and the very child which she had angrily expelled from her presence was recalled, and all the treasures of her maternal tenderness poured out upon him:

"A cat, long an inmate of this house, kittened this spring, and one of her offspring, a Tom, being given her to rear, she proved a most fond and solicitous mother. The kitten grew and throve, and soon became a very fine and playful young cat. The maternal feelings were constantly developed, the mother calling it, licking it, sharing and promoting its frolics, and exhibiting the tenderest anxiety and jealousy whenever any strange person approached.

"In the midst of this exuberant affection a change passed over the cat, and the young one suddenly became the object of hate and irritation to the formerly loving mother. She would not allow it to approach her; and if it only dared to look at her, she would spit and hiss and fly at it, becoming absolutely savage when she found it near her.

"It soon became evident that there would be another litter of kittens, and this sudden change of manner was probably instinctive on the part of the cat, who found herself unable to join in the usual gambols.

"One day, however, a second revulsion of feeling took place: she called her first-born in the most tender and yearning tones, and tried to entice it up-stairs with her. She was so anxious to have her son with her, that she even tried to drag him up-stairs by the neck as she used to do when he was a little kitten.

"Two days afterward the second family was born, and all of them met a watery death. The cat did not seem to miss or regret her lost young, but took back her first-born in their place. Though as large as its mother, it at once resumed all the habits of its infancy, sucking as it had been accustomed to do. The mother licked and caressed it, just as if it had been a new-born kitten, and displayed the greatest anxiety when the postman or any stranger approached. The young Tom still continues to suck, though he has caught many mice and eaten them."

A very similar event occurred lately (1873) in my own house. My cat, called by the children "Duckie," had a family, out of which two were saved. These grew to be cats, and, in the ordinary course of events, were sent off by their mother. In the mean while a new family arrived, but, as we already had three cats in the house, they were at once dismissed from a world in which there was no place for them. Their mother immediately took the two former kittens into favor; and the oddest thing was that she treated them exactly as if they had been tiny helpless kittens a few days old.

Her conduct reminded me very much of that which we often see in parents, especially if they live with or near their children. They really can not understand that a man of forty or a woman of thirty are any thing more than children, and are greatly discomposed whenever these elderly children venture to think or act for themselves. It is the same with old servants; and there are many parents of large families who to the old nurse remain "Master Tommy" or "Miss Emily" to the end of the chapter.

The next anecdote relates to the dog, and shows that in a civilized dog, so to speak, parental affection can endure as in a civilized human being:

"My dog 'Georgie' (short for Georgina) has a daughter, named 'Poppie,' whose father was a collie, she herself being a retriever. People said that it was not safe to keep a mongrel of that description, but experience has proved the mistake.

"She is now (1873) five years old, and the affection which exists between mother and daughter is really beautiful. They always sit close together, and Georgie playfully pinches her daughter all over. If they have been separated by any chance, the daughter comes up wagging her tail, and then licks her mother's nose and mouth.

"Sometimes they go out rabbit-hunting together, and always act in concert. Each of them takes an opposite side of a whin-bush, and one keeps watch while the other scrapes. They perfectly comprehend the meaning of each whine or bark, and no two ladies could understand their own language better than did these dogs, or be more companionable to each other."

Here is also another proof of the fact that animals have a language of their own by means of which they can convey definite ideas to each other, nearly if not quite as well as we can do with the aid of words.

One of the most beautiful characteristics of parental love is the utter abnegation of self which it gives. This is chiefly shown when the young are in danger. A human mother in charge of her child will defy a danger before which she would shrink if alone, and in defense of her offspring will dare deeds of which most strong men would be incapable. For the time her selfhood is extinguished, and her very being is merged into that of the child; and rather than a hair of that child's head should be touched, she would calmly consent to endure the worst tortures that could be inflicted upon her. Indeed, if she would not do so, she would be no true mother, and would degrade herself below the beasts and the birds, who have no hesitation in performing that duty to their offspring, though *savans* do say that they only possess "storgë," whatever they may mean by it, and not parental love. I will now give a few instances of the marvelous courage inspired by parental love in the lower animals.

Every one who has paid even a passing attention to the habits of birds must have noticed that the spotted fly-catcher has a habit of selecting some favorite perch, which it frequents from day to day, scarcely ever changing its haunts. From its coign of vantage it keeps anxious watch around, and when it sees an insect on the wing, dashes off, captures it, and returns with its prey to its perch. It may possibly catch insects when they are not on the wing, but I never knew a fly-catcher do so.

In my garden there is a young mulberry-tree, which is highly prized, having been sent specially from Japan, and being the only survivor of six, the others being all killed by nocturnal cats, who found the stems exactly suitable for sharpening their claws. Of course the young tree was watched with exceeding care, and it was soon seen to become the favorite perch of a spotted fly-catcher.

The bird followed the usual customs of its kin, but after a while it began to act in a very strange manner, fluttering backward and forward between the house and the tree, chirping in a loud and distressed tone, and evidently possessed by anger as well as fear. The cause of its extraordinary action was soon seen to be a cat, which was crouching in front of the ventilation-aperture of the ground-floor, and apparently watching something behind the bars. The bird tried in vain to draw off the cat's attention, fluttering so closely that I feared lest pussy should strike it down, and even at times pecking at the animal's tail.

On removing the cat, a young bird was seen within the grating, evidently the offspring of the fly-catcher. These birds have a way of building their nests in very odd places, and I surmise that in the present case the parents must have made their way through a hole under the steps, and so have reached the ventilating-shaft.

As soon as the cat had been removed, the mother-bird, regardless of my presence, flew to the grating and began to feed the young one. She then went off to a little distance and called her offspring. The poor little bird did all in its power to get through the bars, fluttering its wings and answering its mother with piteous chirps. I felt quite uneasy about them both, for the cat was sure to come back again, and the mother was so bold and reckless in her assaults that I feared for her life; and if she had been killed, the young one must have died of hunger.

So I tried to see if the young bird was sufficiently fledged to use its wings, as in that case it might be let out; but it was so timid that it retreated into the darkness as soon as I approached, and would not let me examine it. An opera-glass, however, overcame the difficulty, and, finding that the young bird was fully fledged, I cut away one of the bars so as to leave a passage, and went to some little distance.

The mother, who was anxiously watching me from the roof of an outbuilding, went at once to the spot, and, after much calling, induced her offspring to come out of the aperture which had been made for it. The delight of the two was beautiful to see; but the mother evidently had the cat in her mind, and did not mean to waste any time in placing her child in safety. So she induced it by degrees to follow her up the branches of an apricot-tree, and thence to the roof of the house, where even a cat could not follow.

In his "Birds of Ireland," vol. i., page 115, Thompson relates an anecdote of a spotted fly-catcher. It had chosen for its resting-place the unglazed window of an outhouse at Beechmount, and had there built a nest, "which was so composed of cobwebs inside and outside that no other material was visible. From its choice of this fragile building substance, the spotted fly-catcher is called 'cobweb-bird' in some parts of England. On the nest alluded to being approached, when it contained young, the parent-bird was very bold, flying angrily at the intruder, uttering shrill cries, and approaching him so near that it might have almost been struck with his hand."

The same writer mentions that the spotted fly-catcher is equally bold toward other birds, beating away all which dare to approach their nest. It is perhaps worthy of notice that, in the instance which I myself observed, I did not once see the male bird; possibly he may have fallen a victim to the cat.

The swallow is equally courageous in defense of her nest. Some little time ago there was a swallow's nest in the porch of the rectory at Adisham—the bird being, of course, carefully protected. Not knowing of the nest, I happened to be standing near the porch, and was much annoyed by a swallow, which persisted in flying round and round, uttering its shrill, screaming cries, and occasionally darting close to my face. It was not until some little time had elapsed that I suspected the cause of the bird's behavior, and then, on looking around, saw the nest and the young in an angle of the porch.

In all these cases the bird had no hesitation in matching itself against foes from which it would have shrunk in terror had not the love of offspring overpowered the love of life. It does not in the least matter what the foe may be, the parent attacking the most powerful enemy with as little hesitation as if the relative proportions of size and strength were reversed. A snake, for example, is specially feared by birds, especially if it be a venomous one; and yet, if a snake threaten the nest of a bird, she will not hesitate to attack as fiercely as if the poison-fangs belonged to her, and not to her foe. The following account, published in the *Dumfries Courier*, 1853, shows how completely parental love will overcome fear, and will induce a feeble bird to fiercely attack a creature from which she would have fled but for the supreme power of love:

"While Mr. Charles Newall, granite-hewer in Dalbeattie, was plying his vocation at Craignar quarry, his attention was suddenly arrested by cries strongly indicative of distress, proceeding from one or other of the feathered denizens of the wood.

"On throwing from him his tools, and hurrying to the spot whence the sounds proceeded, he discovered a robin, apparently in a state of the greatest agitation, whose movements immediately certified him of the true cause of alarm. An adder, twenty inches long and one inch in circumference, had managed to drag itself up the face of the quarry, and was at that moment in the very act of protruding its ugly head over the edge of a nest, built among the stumps of cut-down bushwood, which contained poor Mother Robin's fledged offspring.

"Her maternal instinct prompted her to the only defense of which she was capable. She was engaged, when Mr. Newall first got his eye on her, in alternately coming down, the one moment upon her spoliator, darting her beak into its forehead, and anon rising on the wing to the height of a yard or so above the scene of danger. It was the act of a moment for Mr. Newall to dislodge the aggressor; but in doing so two of the little birds were thrown out of their nest, where, however, they were speedily and carefully replaced.

"While Mr. Newall was engaged in killing the adder, the joy of the parent-bird was so excessive that she actually perched on the left arm of her benefactor, and watched with an unmistakable and intense delight every blow inflicted by his right arm upon her merciless and disappointed enemy; and when that enemy was dead, she alighted upon and pecked the lifeless trunk with all her vigor. Revenge thus taken, she entered her nest, and, having ascertained that all was safe, swiftly repaired to a neighboring branch and piped, as best she could, what was no doubt meant for a hymn of gratitude and a song of triumph.

"When at work since, Mr. Newall has been evidently recognized by the tiny biped; and we do hope that nothing may occur to interrupt a friendship originating in circumstances so specially interesting."

In this account we have several characteristics common to man and the lower animals. First, there is parental affection; next there is courage emanating from that effect; then there is reason, which told the bird that the man, whom it would have regarded as an enemy but for his attack on the snake, was really a friend; lastly, there is revenge, inducing the bird to peck at the body of its dead foe, just as a savage insults and mutilates the carcass of a slain enemy.

From the description of the snake which is here given, it is tolerably evident that the reptile was a viper, those creatures having a special habit of climbing trees and robbing birds of their young. It has often been found by those who have killed vipers that, after receiving the first stroke, the reptile has opened its mouth and disgorged several young birds, in order to lighten itself and enable it to escape more quickly. The common, harmless snake, sometimes called the "grass-snake," mostly contents itself with frogs.

The preceding anecdotes show active courage in the parent; but in Hardwicke's *Science Gossip* for 1873, p. 204, there is an interesting account of passive courage in a partridge:

"The affection and solicitude of the female partridge for her young is very great, and instances are frequently seen by the rural naturalist in his rambles. The closeness with which she will sit when about hatching is remarkable. I once found a nest containing seventeen eggs, on which the female was sitting, and, instead of flying rapidly away when I approached, she allowed me to stroke her glossy head and soft plumage, seeming to appreciate the familiarity. Her confidence gained its reward, as all of the eggs were duly hatched.

"A gentleman in this neighborhood, when jumping across a hedge, alighted with a foot on each side of a partridge-nest, where the female was sitting. The affectionate bird did not stir, even allowing the gentleman to stroke and fondle her. But more admirable still is the address with which both male and female will draw the spectator away from the neighborhood of their brood. Last July, when walking along the highway, I disturbed two partridges near some tall grass. With startled cries they whirred away; and, alighting a few yards off in the middle of the road, went through a series of manœuvres, as if desperately wounded, both of them groveling along on their bellies in the dust, and seeming to tumble over and over in their eagerness. Stopping some distance off, they began to utter curious plaintive cries.

"Being somewhat in a hurry, I did not institute a search for the cause of this little drama, the young; but I have seen a similar instance, in which case I captured one of the plump little chicks, and held it for a time in my hands; but the distress of the old bird became so great that I soon released it. In June, 1868, a pair of partridges had their nest in the clover field opposite, the mowers thoughtfully leaving a tuft of clover to shield the nest. It was very amusing to see how careful the old birds were to prevent attention being drawn to their almost exposed nest. Both of them would go in search of food, and then fly back into the field together; alighting within a few yards of the nest, and having anxiously scanned the neighborhood for a time, the female would slyly approach in a crouching attitude, and creep into the nest."

The proverbial skill of the lapwing in feigning lameness is too familiar to need description.

It is, perhaps, scarcely necessary to cite here the shamefully cruel plan that was formerly used by whalers to secure their prey. If they met with a young whale, or "calf," as they called it, they always used to harpoon it, knowing that its mother would come to its rescue, and be so regardless of her own safety that there was neither difficulty nor danger in harpooning her also.

I believe that this atrocious custom is now abandoned, though I fear from commercial rather than conscientious motives. The calf is all but useless; whereas, if it were allowed to live, it would grow into a whale, and fill sundry barrels that would otherwise have to go home empty. The fact, however, remains that the whale is so utterly forgetful of self, when its offspring is in danger, that it neglects its usual wary habits, and so falls a victim to parental affection.

The following curious story of parental affection was communicated to me by a lady expressly for this work:

"Some years ago (in 1868) our steward and his wife lived in the lodge at our east gate, distant about half a mile from the house. As a favor, the steward's wife allowed a common duck to sit upon a number of duck's eggs, which, according to agreement, were to be taken away as soon as they were hatched. In due time the eggs were hatched, and the young ducklings removed to our house, and placed in the poultry-yard, under the charge of a hen who had already a few ducklings to look after. The yard in question is protected by a wall nearly three feet high, and upon this wall is a wire netting seven feet in height.

"In the afternoon of the same day the mother duck (who had never left the lodge in her life) came waddling up all the way to the stables, got on the top of the wall, and managed to get her own little ducklings through the wire netting. Having done this, she took them back to the lodge, leaving the hen in quiet possession of the ducklings of which she previously had charge.

"As soon as this was known, the ducklings were again taken to the house, and the duck shut up in a dark place at the lodge for two days. But no sooner did she gain her liberty than she made another journey to the poultry-yard, and again began to drag her offspring through the wire net, this time killing one by letting it fall off the dike. I therefore bought the duck from her owner, placed her in the poultry-yard, and allowed her to bring up her brood peacefully in her own way."

How did this duck find out her young? I imagine that it must have been by the sense of hearing. Ducklings, when separated from their mother, or when lost, always make a considerable outcry. The mother duck had probably heard them crying, had gone off in the same direction, and when she got near the poultry-yard had been directed by their voices. It is, moreover, evident that both the mother and the children must have understood each other's language, as by no other means could she have called her young brood to the fence, and directed them to remain there while she pulled them through one by one.

It really seems, in this as in many other instances, as if, in their own language, the animals had names known to themselves, and the Robin, Dicky, Flapsy, and Pecksy of fiction to be not so much fictitious as we might fancy. In feeding their young, birds always take them in their proper turn, and how they can do so without some means of calling them by name, especially in the case of birds which hatch many eggs in each brood, is more than I can understand. Both birds and animals know and answer to names given to them by man in human language, and I see no reason why they should not equally know and answer to names given by themselves in their own language.

I may here mention that the love of a bird for the young which she hatches has always been somewhat of a problem to me. In the case of the mammalia, there is no difficulty in understanding that the mother should feel love for the creature who is absolutely part of herself —whose very life-blood is drawn from her veins. But this is not necessarily the case with birds. If, as it often happens with poultry, the eggs of several hens are placed under one bird for hatching, the hen who hatches them knows no difference between the chickens that proceed from her own eggs and those which are developed from the eggs laid by others.

This curious trait of character holds good even where the eggs belong to birds of different species. Take, for example, the very common instance of a brood of ducklings being hatched and reared by a barn-door hen. The hen displays as much affection for the young ducklings as if they had proceeded from her own eggs, and this in spite of the disparity of instinct and habit, which becomes stronger in proportion to the ducklings' growth.

May it not be that parental love may have different channels of transmission, and that in such a case as this the emanation from the sitting hen may be the vehicle of parental love toward the young which are to be hatched? Certain it is to those who observe that a sitting hen is altogether a changed being, both in attitude and expression. She is entirely absorbed in the eggs which she is incubating, and, though she may not have intellect enough to distinguish a plaster-of-paris imitation or a mere lump of chalk from one of her own eggs, love is independent of intellect, and may exist in all its strength, though it may be wasted on an unworthy object.

As I have already remarked, under the heading of "Conjugal Love," fishes are not particularly emotional beings, and are not likely to entertain a lasting love for any thing. Indeed, in some cases parental love would be absolutely useless, as in the case of the cod-fish, which could hardly be expected to entertain a special love for each of the countless thousands of young which it produces every year. At least, if such were the lot of the mother, her life would be any thing but enviable, considering the varied foes that beset her eggs as soon as they are produced, and her young as soon as they are hatched.

Just, however, as there are fishes which possess conjugal love, so there are fishes which possess parental love, and the chief of these is the stickleback. Many accounts have been written of the proceedings of this remarkable fish, but the best that I have seen was written by the late J. Keast Lord, in his "Naturalist in British Columbia." And the curious point is that parental love in the case of the stickleback belongs to the father, and not to the mother. Indeed, as there is one father and a considerable number of mothers, it is the only arrangement that could be made.

Inverting the usual order of things, the whole labor of providing for the young, which is very considerable, devolves upon the male, the female doing nothing except lay her eggs, and let the male look after them.

Mr. Lord's description of his proceedings must be given in his own words:

"I have often, when tired, lain down on the bank of a stream beneath the friendly shade of some leafy tree, and, gazing into its depths, watched the sticklebacks either guarding their nests already built or busy in their construction. The site is generally among the stems of aquatic plants, where the water always flows, but not too swiftly. He first begins by carrying small bits of green material, which he nips off the stalks, and tugs from out of the bottom and sides of the banks. These he attaches by some glutinous material, that he clearly has the power of secreting, to the different stems destined as pillars for his building.

"During this operation he swims against the work already done, splashes about, and seems to test its durability and strength; rubs himself against the tiny kind of platform, scrapes the slimy mucus from his sides to mix with and act as mortar for his vegetable bricks. Then he thrusts his nose into the sand at the bottom, and, bringing a mouthful, scatters it over the foundation. This is repeated until enough has been thrown on to weight the slender fabric down, and give it substance and stability. Then more twists, turns, and splashings, to test the firm adherence of all the materials that are intended to constitute the foundation of the house that has yet to be erected on it.

"The nest, or nursery, when completed, is a hollow, somewhat rounded, barrel-shaped structure, worked together much in the same way as the platform fastened to the water-plants; the whole firmly glued together by the viscous secretion scraped from the body. The inside is made as smooth as possible by a kind of plastering system; the little architect continually goes in, then, turning round and round, works the mucus from his body onto the inner sides of the nest, where it hardens like a tough varnish. There are two apertures, smooth and symmetrical as the hole leading into a wren's nest, and not unlike it."

I have seen plenty of these little nests, and always regretted the extreme difficulty of preserving such beautiful specimens of fish-architecture. Unfortunately, although they answer very well as long as they are under water, they do not hold together when removed into the air, the peculiar cement not being sufficiently strong to bear the unsupported weight of the materials.

Having thus prepared his house, the fish sets off in search of a partner to grace it. This she does but for a very short time, simply passing in at one aperture and out at the other, remaining some five minutes in the nest, and during that time depositing her eggs. Having finished, she passes out, followed by the male, who goes and brings another female, and repeats this process until the nest is furnished with as many eggs as it can hold.

He then places himself on guard, and watches his treasure as vigilantly and fiercely as a tigress watches her cubs. He often has to fight hard battles, for there is no delicacy so loved by fish as the roe of other fish, even of their own species; and the nest is sure to be beset by sticklebacks or other fish, and water-beetles, trying to get at the eggs. For some six weeks he keeps this anxious watch, and even when the young are hatched he does not desert his post. It is said that he will not allow them to wander far from the nest, and that if one of them should stray beyond certain limits, he will seize it and bring it back again.

In the encounters which he has to undertake he runs much risk of losing his life, for the sharp spines with which the body is armed are weapons which can be used with fatal dexterity. Each fish tries to force its way under the other, and, if it can succeed, rises rapidly, and drives the spines into the sides or belly of its adversary, often causing its death, and always wounding it seriously. Even in fishes, then, we see parental love sufficiently developed to induce the male stickleback to remain for six weeks on guard, to fight any foe that may attempt to rob him of his treasure, and to risk and sometimes to lose his life in the defense of his offspring.

The reader will not fail to have noted the curious fact that, whereas parental love is, in nearly all creatures, chiefly manifested in the mother, in this case the mother never troubles herself about the fate of the eggs which she has deposited, but leaves them all to the father. Neither does she take any share in the preparation of the nest, the whole of the labor belonging to the male, who has to gather materials, make the nest, get it stocked with eggs, guard it at the risk of his life, and see the young safely started in life. The human parallel is too obvious to need mention.

There are other fishes in which the male takes the chief part in the incubation of the eggs. Such, for example, are the curious Lophobranchiate fishes, of which the common bill-fish, or pipe-fish, and the quaint little sea-horse are good examples. The former, by the way, is much more plentiful than is generally supposed, and I have found many of them served up among the tiny fishes which are called by the general name of whitebait. In all these fishes the males are provided with some special apparatus, such as a pouch, a double ridge of skin, etc., by means of which the eggs are attached to the body of the male until they are hatched.

Then there are certain birds, mostly belonging to the Cursoria, or that group of which the ostrich is the type, the females of which take no trouble about their eggs after laying them, but depute the whole of that business to their mates.

An instance where a spider defended its eggs against most formidable enemies is narrated by Mr. F. C. Rawlins in Hardwicke's *Science Gossip* for April, 1873.

In a recent number I saw some interesting matter relative to spiders and their poisoning apparatus. The following, which comes from personal observation, will vouch for the efficacy of this apparatus, and also show what a weapon of defense it becomes when the parental instinct is roused by an attack upon the offspring:

"One day in the autumn I captured a fine specimen of the garden spider (*Epeira diadema*), which was running over a flower border, skillfully conveying the precious filmy bag of eggs underneath its body over the various obstacles which impeded its progress. It did not seem averse to the shelter afforded by a small wooden box, and remained at one end with its treasure so contentedly that I left it for a few moments, and placed it on the top of a dahlia-pole.

"On returning I discovered that an exploring party, consisting of four ants, was scaling the walls of the fortress. Until they were fairly within its walls the spider seemed unaware of their approach; and, in fact, until a forcible attempt was made by the intruders to grapple with the egg-bag, it remained strangely apathetic. But this insult offered to the helpless young was too much. It darted forward and assailed the foremost. It was a tough fight—four to one—but the valiant mother conquered in the end; for three of the invading foe lay dead (evidently poisoned by a venomous bite), and the fourth was fairly driven off. The victor then retired with her insulted property to a corner, and I carried off the box.

"An untimely escape prevented an experiment I hoped to make, viz., of trying to tame this member of the usually disliked Arachnida family."

It would have been easy enough to fill the whole of the book with stories of true parental love among the lower animals, but I have selected these in order to show that the feeling is identical in man and the lower animals, although, of course, the mode of manifesting it must differ. First, we see the untruth of the theory that parental love is life-enduring in man and only brief among the animals. We see that, in proportion to the duration of life, it is quite as brief among the savages as among the animals. Then we have examples where parental love has been lost and then restored, and also where it never was lost.

We see how, in the animals as well as in man, parental love causes complete abnegation of self, the parents living for their children, and not for themselves. We see how it gives strength to the weak and courage to the timid; that even the very fishes and the spiders are amenable to the same influences as man, and that parental love, one of the highest and holiest feelings of which a living and immortal soul can be capable, is shared equally by man and beast, according to their respective capacities.

CHAPTER XVII.

THE FUTURE STATE.

Immortality of Man as Treated in the Old Testament.—The Lower Animals not Thought Unworthy of a Divine Law.—Man and Beast Equally Liable to Punishment for the Same Crime.—Instinctive Belief in Immortality.—The Spirit of Animals Developed by Communion with the Spirit of Man.—Opinions of Various Writers on the Subject.—Eugenie de Guérin and Mrs. Somerville.—The Contemplative and Logical Minds.—Southey's Epitaph on his Dog.—Lamartine on a Similar Subject.—The Doctrine of Apparent Inequality and Compensation.—How to Reconcile Pain and Suffering with Divine Justice.—The Different Lots of Man and Beast.—The Object of Suffering.—Individuality connected with Immortality.—Individuality often Overlooked, even though it be Strongly Marked to those who can Detect it.—The Groom and the Engine-driver.—Individuality Retained in the Next World, and Developed there.—Mr. J. Nelson Smith on a Dead Lion.—The Spirit of the Beast; Comparison with that of Man.—Death and its Results in Man and Beast.—Spirit and Matter.—The Dead and the Living.—The Spiritual and Material Eye.—The Story of Balaam.—The Cat and the Apparition.—Parallel of the Two Narratives.—Epilogue.

I HAVE already shown, at the beginning of this work, that, contrary to the popular tradition, the Scriptures do not deny a future life to the lower animals. We will now see if Scripture has any thing to say in favor of another world for beast as well as man.

It is a very remarkable point that even as to the immortality of man, the Scriptures of the Old Testament teach that doctrine rather by inference than by direct assertion.

I presume the reason to be that the writers of the various books, which were at a comparatively late period selected from among many others and formed into the volume which we popularly call the Bible, assumed as a matter of course that man was immortal, and did not trouble themselves to assert that which they supposed every one to know already.

As far as the Old Testament goes, inference tells much stronger in favor of the beast's immortality than in that of man; for although in either case there is no definite assertion of a future life, there is at all events no such denial of the immortality of the beast as we have seen to be the case with man (see page 10).

We all know that the beasts were included in the merciful provisions of the Sabbath, which was in its essence a spiritual and not merely a physical ordinance. Then we find in the ancient Scriptures many provisions against maltreating the lower animals, or giving them needless pain; and these provisions occupy an equal place in the Divine Law with those which treat of man.

See, for example, the well-known prohibition of "seething a kid in its mother's milk," this being apparently some cruel heathen custom during harvest-tide. Then the ox which is used in treading out the corn is not to be muzzled, lest it should suffer the pangs of hunger in the presence of the food which it may not eat. Even such a trivial custom as bird's-nesting is regulated by Divine Law (see Deut. xxii. 6, 7). As, moreover, many animals must be killed daily, some for sacrifice and others only for food, the strictest regulations are given that their mode of death shall be sharp and swift, and that the whole of their blood shall be poured out upon the earth, thus preventing them from lingering in pain.

I need scarcely refer to the last few sentences of the Book of Jonah:

"Thou hast had pity on the gourd, for the which thou hast not labored, neither madest it grow; which came up in a night, and perished in a night:

"And should I not spare Nineveh, that great city, wherein are more than sixscore thousand persons that can not discern between their right hand and their left hand; *and also much cattle?*"

Again, see Psalm l. 10, 11:

"Every beast of the forest is mine, and the cattle upon a thousand hills.

"I know all the fowls of the mountains: and the wild beasts of the field are mine."

The Scriptures are full of similar passages, in which God announces himself as the protector of beast as well as of man; among which we may reckon the well-known saying of our Lord respecting the lives of the sparrows.

Allusion to this branch of the subject is made by Cowper in his "Task:"

"Man may dismiss compassion from his heart,
But God will never. When He charged the Jew
To assist his foe's down-fallen beast to rise;
And when the bush-exploring boy, that seized
The young, to let the parent-bird go free;
Proved He not plainly that His meaner works
Are yet His care, and have an interest all—
All in the universal Father's love."
COWPER'S *Task*.

There is, however, one passage which certainly does seem to point to a future for the beast as well as for man, and does at all events place them both on a similar level. It occurs in Genesis ix. 5, and forms part of the concise law which was delivered to Noah, and which was the forerunner of the fuller law afterward given through Moses: "Surely your blood of your lives will I require: at the hand of every beast will I require it, and at the hand of every man; at the hand of every man's brother will I require the life of man."

And this injunction was afterward incorporated into the Mosaic law, where an ox who kills a man is subject to death, just as if it had been a man who had murdered one of his fellows (see Exodus xxi. 28).

As a writer in the *London Review* well said, some years ago: "There would be no meaning in this retribution if the animal had no living soul to be forfeited, as the human soul had been yielded to death. It would be absurd to destroy a vegetable which had caused the death of a human being, inasmuch as it has no soul. It was not considered absurd to destroy an animal under such circumstances, inasmuch as it has a soul."

Thus, while there are no passages of Scripture which deny the immortality of the lower animals, there are some which certainly tend toward inferring it; but I do not see how we could expect to gain any information on the subject from the Scriptures, which were written for human beings, and not for the lower animals; and, as we find so few direct references to the future state of man, we could hardly expect to receive direct instruction upon that of beasts.

But just as man has always had within himself an intuitive witness to his own immortality, so do I find that all who have watched the ways of the lower animals have possessed an instinctive sense that they too must have a future life. Some, it is true, have been led away by a wrong interpretation of the passages in the Psalms, Job, and Ecclesiastes; but, in conversing with them, I have always found that underlying this idea is a feeling that animals, which surpass many human beings in love, unselfishness, generosity, conscience, and self-sacrifice, must share, together with those virtues, an immortal spirit in which they take their rise.

For myself, I attribute to the conduct of my dog "Rory" my firm conviction that for such animals a future life must be in store; and if for him, why not for all? It is true that in him the moral sense of duty was developed to a very high degree, as were his reasoning powers and the faculty of love. I could not believe that an animal which would die of grief, as he died, for the absence of his master, could have his existence limited to this present world, and that such intensity of love should terminate at the same moment that the material heart ceased to beat.

But, though in his case these higher qualities were so greatly developed by constant communion with a human spirit, there are hundreds of thousands of dogs with similar capabilities, but without similar advantages. I feel sure that they will have the opportunity of developing their latent faculties in the next world, though their free scope has been denied to them in the short time of their existence in this present world.

I have been rather surprised to find how many standard writers have held these opinions. All students of theology are acquainted with the passage in Bishop Butler's "Analogy" in which he states that the Scriptures give no reasons why the lower animals should not possess immortal souls. I will now take passages by two very celebrated women, the former a representative of devotional religion, and the latter a thorough mistress of the physical sciences, and a deep mathematician.

The first extract is taken from the "Diary" of Eugénie de Guérin, and is a remarkable instance of the manner in which the contemplative human soul yearns after companionship with the souls of fellow-creatures that have been loved and have passed away:

"*1st August*, 1835.—This evening my turtle-dove has died; I know not from what cause, for it continued to coo up to to-day. Poor little creature, what regret it causes me! I loved it; it was white; and every morning it was the first voice I heard under my window, in winter as well as in summer. Was it mourning or joy? I know not, but its songs gave me pleasure. Now I have a pleasure the less: thus each day we lose some enjoyment.

"I mean to put my dove under a rose-bush on the terrace: it seems to me that it will be well there, and that its soul (if soul there be) will repose there sweetly in that nest beneath the flowers. I have a tolerably strong belief in the souls of animals, and I should even like there to be a little paradise for the good and gentle, like turtle-doves, dogs, and lambs. But what to do with wolves and other wicked minds? To damn them?—that embarrasses me."

She might have reflected that, in its place, the wolf is as useful and as innocent as the lamb. It has an object in life, and carries it out until that object be attained, when it perishes, as has been the case in our own country, not only with the wolf, but with the bear and other predacious animals.

The next passage is taken from Mrs. Somerville's "Memoirs." I have selected these two because they represent two differently constituted minds, which yet agree just on the very subject where one would have expected the greatest divergence.

The one is essentially devotional, trusting to intuitive ideas, and not having the least pretense to logic, or even a sequence of reasoning. The other is a mind trained to observation, to mathematical accuracy, to hard reasoning, and to that faculty which is so seldom seen in the female sex —namely, the power of generalization. Speaking of death, and the accompanying change of surrounding objects, Mrs. Somerville, then aged eighty-nine, proceeds as follows:

"I shall regret the sky, the sea, with all the changes of their beautiful coloring; the earth, with its verdure and flowers; but far more shall I grieve to leave animals who have followed our steps affectionately for years, without knowing for certainty their ultimate fate, though I firmly believe that the living principle is never extinguished. Since the atoms of matter are indestructible, as far as we know, it is difficult to believe that the spark which gives to their union life, memory, affection, intelligence, and fidelity is evanescent.

"Every atom in the human frame, as well as in that of animals, undergoes a periodical change by continual waste and renovation: the abode is changed, not its inhabitant. If animals have no future, the existence of many is most wretched. Multitudes are starved, cruelly beaten, and loaded during life; many die under a barbarous vivisection.

"I can not believe that any creature was created for uncompensated misery: it would be contrary to the attribute of God's mercy and justice. I am sincerely happy to find that I am not the only believer in the immortality of the lower animals."

We will presently revert to the latter part of this interesting letter. I can not but notice the remarkable fact that two minds so differently constituted should have arrived at the same result in two different ways. The one does not pretend to any process of reason, but passes at once, *per saltum* as it were, to the firm belief that the lower animals must have a future life. The other works her way to the same point through a consecutive train of reasoning, basing her arguments upon physical facts of which Madame de Guérin was entirely ignorant. We instinctively agree with the one, and we can not disagree with the other.

Having now seen the manner in which the contemplative and logical female minds treat this subject, let us turn to the masculine mind. We will take for example Southey, a man of singularly deep and wide reading, possessed of the exceptional gift of rendering poetical the least beautiful of subjects. If ever there were a clumsy and repulsive idealization in the world, it may be found in the many-headed and many-armed deities of Hindoo mythology; and yet, in the hands of Southey, they are invested with a glamour like that which Scott threw over the most prosaic and commonplace of landscapes in his native land.

Writing of the death of a favorite spaniel who had been his companion in boyhood, Southey proceeds as follows:

"Ah, poor companion! when thou followedst last
Thy master's parting footsteps to the gate
Which closed forever on him, thou didst lose
Thy best friend, and none was left to plead
For the old age of brute fidelity.
But fare thee well. Mine is no narrow creed;
And He who gave thee being did not frame
The mystery of Life to be the sport
Of merciless man. There is another world
For all that live and move—a better one!
Where the proud bipeds, who would fain confine
Infinite goodness to the little bounds
Of their own charity, may envy thee."

The following extract is taken from "Jocelyn's Episode, par A. de Lamartine," and is translated by the author of "Episodes of Insect Life:"

"My dog! the difference between thee and me
Knows only our Creator:—only He
Can number the degrees in being's scale
Between thy instinctive lamp, ne'er known to fail,
And that less steady light of brighter ray,
The soul which animates thy master's clay;
And He alone can tell by what fond tie,
My look thy life—my death, thy sign to die.
Howe'er this be, the human heart bereaved,
In thy affection owns a boon received;
Nor e'er, fond creature, prostrate on the ground,
Could my foot spurn thee or my accents wound.
No! never, never, my poor humble friend,
Could I by act or word thy love offend;
Too much in thee I reverence that Power
Which formed us both for our appointed hour;
That Hand which links, by a fraternal tie,
The meanest of His creatures with the high.
Oh, my poor Fido! when thy speaking face,
Upturned to mine, of words supplies the place;

When, seated by my bed, the slightest moan
That breaks my troubled sleep disturbs thine own;
When noting in my heavy eye the care
That clouds my brow, thou seek'st its meaning there.
And then, as if to chase that care away,
My pendant hand dost gently gnaw in play;
When, as in some clear mirror, I descry
My joys and griefs reflected in thine eye:
When tokens such as these thy reason speak
(Reason which with thy love compared is weak),
I can not, will not, deem thee a deceiving,
Illusive mockery of human feeling,
A body organized, by fond caress
Warmed into seeming tenderness;
A mere automaton, on which our love
Plays, as on puppets, when their wires we move.
No! when that feeling quits thy glazing eye,
'Twill live in some blest world beyond the sky.

* * * * *

No! God will never quench His spark divine,
Whether within some glorious orb it shine,
Or lighten up the spaniel's tender gaze,
Who leads his poor blind master through the maze
Of this dark world; and, when that task is o'er,
Sleeps on his humble grave, to wake no more."

We will now revert for a time to the subject mentioned at the end of the extract from Mrs. Somerville's "Memoirs." Every one must at some time or another have been struck with the problem of apparent inequality in the lives both of man and beast. We see some human beings endowed with every thing that man can desire—health, strength, wealth, accomplishments, and capacity of enjoyment; while others are destitute of all these accessories to happiness.

Putting aside the fact that some whose lots seem to be the most enviable are the least to be envied, we acknowledge that this inequality does exist, and that the earthly lot of some is very hard, while that of others is very easy. But we are taught in the New Testament the great doctrine of Compensation, which is, in fact, nothing more than justice.

As St. Paul remarks, who spoke from personal experience, the sufferings of this present world are not to be compared with the glories of the world to come; and that, in fact, suffering is the precursor of glory. That some such principle of divine justice must exist was instinctively known long before it was thus explicitly declared. We find references to such compensation throughout the Psalms, in passages too numerous and too familiar to need quotation; and even Job himself, sunk in the very depth of afflictions, could say, "Though He slay me, yet will I trust in Him. . . . He also shall be my salvation" (Job xiii. 15, 16).

As far, therefore, as man is concerned, the problem of apparent inequality is not so difficult of solution. Expectant of a future life, we look forward to it in our worst earthly sorrows, and feel that when we have passed into that new life we shall receive our reward. Thus, in spite of all apparent inequalities of the human lot in this world, we feel that divine justice will be more than vindicated in the world to come, and that when we enter that world we shall understand and acquiesce in the justice that gave a hard lot upon earth to us and an easy one to others.

But, supposing the lower animals to have no future life, what becomes of divine justice? Even in our own country and in our own day, the cruelties that are perpetrated upon animals are a disgrace to the nation. Bad as they are, however, they are as nothing to the horrors which are seen with absolute unconcern in other countries. But, even in our own land, let us take as an example one of the most ill treated of animals—the donkey.

We will suppose the very likely case of two donkeys of the same age and similar capacities being sold to different masters, both costermongers. One of them treats the animal with kindness, and the other with cruelty. The one urges it to its work by kind words, the other by blows and other forms of bodily torture. The one feeds the animal as liberally as his means will afford, while the other leaves the beast, by whose labor he lives, to forage for himself, and spends in drink the money which ought to have been expended in fodder.

One of these animals lives a long and a happy life, doing his work with eager willingness, loving his master, and being loved by him. The other is soon worn out by hardships, trembles at the very sound of his master's voice, and succumbs at last to pain and starvation. I have purposely placed the more favored animal in a laboring sphere of life, because I am sure that it was formed for labor, and that a properly directed life of work is far happier than the state of a petted, pampered, and idle animal.

Now, supposing that animals have no immortal souls and no future life, it is simply impossible to recognize that the Maker of these two animals can be just. The two contrasted lives indicate an injustice too flagrant for any human being to perpetrate unless wholly deficient in ideas of right and wrong. But supposing them to possess immortal souls, and a future life in which those souls shall be developed to the fullest amount of their capacities, then we can at once reconcile those apparent discrepancies with absolute justice and perfect love. Dealing with the lower animals as with ourselves, the Creator looks to the spiritual world, which is eternal, and not to the material world, which is temporal, and by means of the one instructs and prepares his pupils for the other.

Take the most prominent instance of apparent

inequality and injustice—namely, poverty and wealth. We are gifted with wealth, or it is withheld from us, according to our individual capacities. That which is good for one, as a preparation for the future life, is bad for another, and it is given or withheld accordingly. For example, we all know that when our Lord met the wealthy young man who was proud of his riches, and yet desired to be a disciple, the condition of admission was that he should divest himself of all his wealth, and divide it among the poor.

Many persons have inferred from this order that no one ought to possess wealth. But a little reflection will show that the order in question was not universal, but addressed to a single individual, and to no other. There were many rich men with whom Christ habitually associated, notably Joseph of Arimathea, and yet he never advised them to reduce themselves to poverty.

He knew best what was good for each, and *a fortiori* must he know what is good for animals which exist on a lower and more contracted plane than man. I firmly believe, with St. Paul, that the object of suffering in this present world is that it forms a preparation and introduction to the life to come; and I am perfectly convinced that any creature which is capable of suffering has in that very capacity its passport to the eternal life for which its sufferings are but a preparation.

This brings us to another stage in our argument—namely, the possession of Individuality as connected with Immortality.

As for ourselves, did we not possess individuality we should need no diverseness of management, for all would be treated alike. But we see that in man no two faces are exactly alike, simply because no two souls, of which the countenance is an indication, are alike; and the same will hold good among the lower animals.

Looking, for example, at a flock of sheep, there is no apparent difference between them, and a portrait of any one would equally resemble any other. But the shepherd, if he know his business, will be able to distinguish every sheep separately, and can describe the mental peculiarities of each individual.

Again: one yellow canary looks, to ordinary eyes, just like another yellow canary, while in reality the mental character of each bird is impressed upon its countenance as strongly as are human qualities upon the visage of man. I once had some thirty canaries in a single aviary, and not only knew them all by sight, but could anticipate how each bird would act under certain circumstances.

It is this quality, both in man and beast, that implies a separate treatment for each individual, and becomes a plea for immortality. That I am not alone in this idea is shown by the following letter from a correspondent:

"The difference in character between individuals of the same species is as striking as the differences among men.

"My present dog, though very handsome, is a thorough vulgarian in mind. He prefers bad company, lives by choice in the kitchen, is rude and unmannerly, never barks at a beggar, and delights in a general row or a fight over a bone.

"My former dog, 'Nettle,' was a perfect aristocrat. Nothing would induce him to consort with vulgar people, to enter a kitchen, or descend the area stairs. He perfectly understood the importance attached to a large house and handsome furniture. When we were traveling in the Highlands, and had to put up in any lodgings which we could get, Nettle was perfectly miserable. I remember him at Ballater persistently rushing past our shabby house into one next door, which was handsomely furnished. The lady in occupation disliked dogs; so, after capturing Nettle once or twice, when he had made a raid upon our neighbor's premises, we had to watch him when we neared the house, and bring him by force into our mean quarters. At last we secured handsome lodgings; whereupon Nettle's dignity was soothed, and he never mistook his own abode any more.

"These things seem to be trifles, but it is the observation of such apparent trifles in every creature which I have been able to watch carefully that convinces me more of their separate, individual, spiritual life than even the evidences of great intellect that are occasionally given. Were the beasts but mere animated machines, these distinctive characteristics need not exist."

I may here mention that my own cat "Pret" was equally aristocratic in his notions. Nothing would induce him—not even milk when he was hungry—to put his head into the kitchen, or to enter the house by the servants' door.

To me, the manner in which we ignore individuality in the lower animals is simply astounding. See, for example, how the generality of grooms treat all horses as if they were just so many machines turned out of the same mould, and to be treated just like machines. The "Go ahead," "Stop her," "Back her," of the engineer are represented by the whip or spur of the groom, the jerk or savage pull at the bridle; and the groom has no more idea that he is inflicting pain upon the senses of an immortal fellow-creature than has the engineer of hurting the iron and brass of his engine. Indeed, I fear that, as a rule, the average driver is more merciful to his engine than the average groom to his horse, the former sparing it at the descents, and helping it up the ascents by the accumulated force obtained by the rush down the preceding decline.

We have thus in every species a double kind of individuality: first there is one that is common to the entire species, and next there is one that, in addition to this common characteristic, distinguishes each separate being from its fellows. It is the former of these which makes a species to be what it is, and I am firmly convinced that neither is lost in the future life—that both may be capable of development. Thus, I hold that the dog, the horse, the lion, and the elephant will be in the next world what they are in this. They will be better animals in that world, just as we hope to be better men; but they will not approach us any nearer than they do at present.

I will here quote an eloquent passage from a very remarkable book, which is nearly unknown —namely, "The Science of Sensibility," by Mr. J. Nelson Smith:

"Behold the lion, when he comes forth from his den to seize the prey which his own wants and those of his whelps demand, with flowing mane, steadfast purpose, and paralyzing gleam of eye.... If the voice of lightning is fuller in its volume as it peals over the plains, the vibrating death-knell of the lion is more appalling to both man and beast. If the burning ball of electricity is irresistible, the fatal grasp of the lion is no less fatal to animals; if its flash is more vivid, the angry glare of his eye is more terrible to encounter. The terror of all beasts, and undisputed monarch of the forest, he roams from jungle to jungle and knows no fear.

"But the skill of the hunter sends a bullet through the organs of thought, judgment, and will in that self-reliant head: one terrific bound, one desperate sweep of those huge paws in a vain effort to tear the earth from its centre, and down goes the carcass of that fearful monarch of the forest, stark, by the huge rock on which he has so often gamboled.

"A few spasmodic surges, convulsive tremors, and he stretches himself on the ground, an immovable mass of terrestrial matter. Those gleaming orbs are glazed and sightless, and those terrible limbs are stiffened with the chill of death. Still, even that lifeless frame is an admirable statue of animal force and unquestioned courage, and his slayer approaches even his lifeless corpse with fear, and springs back at the slightest tremor of his departing life.

"What made his voice more terrible than thunder, his spring more fatal than its bolt, and where is it gone?

"Since the departure of the soul, the intelligent motive power which was driven out of that muscular structure by the derangement of the machinery of the mind on which it operated and performed those appalling strains in the great drama of life, that terrible structure of animal life is as harmless as a marble statue, and is soon decomposed by the chemical elements which surround it.

"For an hour after its departure the carcass remains warm and pliable. Every limb is perfect, not a muscle of the body is injured; only the organ of will is unstrung, and the spiritual operator departed. And such an operator! Is his knowledge obliterated? Has a leaden missile annihilated a decree of the Almighty, and decomposed a celestial volition?—or has it only released an immortal soul from the prison-house of a terrestrial body, and given it a passport to the sublime joy of its eternal existence?"

In the last sentence the writer has touched upon the central idea of this book—namely, the possession by animals of an immortal soul. The reader may remark that at page 14 I have cited the important passage of Ecclesiastes, in which a spirit is assigned to the beasts as well as to man. Now the very fact that man can transmit his ideas to the lower animals is a proof that they must possess a spirit which is able to communicate with the spirit of man. When, for example, a man gives an order to his dog, and is obeyed, he affords a proof that both possess spirits, similar in quality, though differing in degree. To give an order to a plant would be useless and absurd, because the plant has no spirit which can respond to the spirit of the man. But the spirit of the dog can and does respond to the spirit of the man—and the two will equally live, each on its proper plane, after the earthly body has been resolved into its elements.

One of our poets has rightly said—

"Man never dies: the body dies from off him;"

and this is equally true of man and beast. The change which we call death is but a more rapid disengagement of the spirit from the body than that which is perpetually taking place. The body is unceasingly separating itself from the spirit, and whether in the waking or sleeping hours the earthly particles which the spirit has accreted around itself are constantly being thrown off. In fact, the death of the body is ever with us, and is a necessary concomitant of the temporary connection between the immortal spirit and the material world.

We now advance one more step.

We all know that spirit can not act directly upon matter, and *vice versâ*. The earthly eye, for example, can not see spiritual objects. But the spiritual eye, which gives force and potency to the optic nerves of the material eye, can do so if the outer veil of flesh be for a while removed. Take, for example, a few instances of such extended vision as given in the Scriptures. First, there is the case of Elisha's servant, whose spiritual eyes were opened, *i. e.*, enabled to pierce through the veil of the flesh, and who was enabled to see the hosts of spiritual beings by whom the place was surrounded. Similarly, when the shepherds saw the angels who announced the birth of Christ, and when the three apostles saw Moses and Elijah, they saw these spiritual beings with the eye of the spirit, and not with that of the flesh.

There are, as we know, many persons who can not believe that, as they put it, the living should be able to see the dead. Neither do I believe it. But as the spirit lives, though the material body no longer inclose it, surely there can be no difficulty in believing that the living spirit within an earthly body may see a living spirit which has escaped from its material garment. We do not doubt that after the death of the body the spirit will live and see other spirits similarly freed from earth, and it is no very great matter that the living should see the living, though one be still enshrined in its earthly tabernacle, and the other released from it.

This being granted—and it is not very much to grant—it necessarily follows that if the lower animals possess spirit, they may be capable of spiritual as well as material vision. That they do possess this power, and that it can be exercised, is shown by the story of Balaam. There we find it definitely stated, not only that the ass saw the angel, but that she saw him long before her master did. Now the angel, being a spiritual being, could only be seen with the spiritual eye; and it therefore follows that, unless the story be completely false, the animal possessed a spirit, and saw with the eye of that spirit.

I should think that none who believe in the truth of the Holy Scriptures (and I again remind the reader that this book is only intended for those who do so), could doubt that here is a case which proves that the spirit of the ass was capable of seeing and fearing the spiritual angel. And if that be granted, I do not see how any one can doubt that the spirit which saw the angel partook of his immortality, just as her outward eye, which saw material objects, partook of their mortality. Shortly afterward the eyes of the prophet were opened, and he also saw the angel; but it must be remembered that the eyes of the beast had been opened first, and that she, her master, and the angel met for the time in the same spiritual plane.

I have for a long time had in my possession a letter from a lady, in which she narrates a personal adventure which has a singularly close resemblance to the Scriptural story of Balaam. It had been told me immediately after I threw out my "feeler" in the "Common Objects of the Country." As I had at that time the intention of vindicating the immortality of the lower animals, I requested the narrator to write it, so that I might possess the statement authenticated in her own handwriting.

At the time of the occurrence, the lady and her mother were living in an old country chateau in France.

"It was during the winter of 18— that one evening I happened to be sitting by the side of a cheerful fire in my bedroom, busily engaged in caressing a favorite cat—the illustrious Lady Catharine, now, alas! no more. She lay in a pensive attitude and a winking state of drowsiness in my lap.

"Although my room might have been without candles, it was perfectly illuminated by the light of the fire. There were two doors—one behind me, leading into an apartment which had been locked for the winter, and another on the opposite side of the room, which communicated with the passage.

"Mamma had not left me many minutes, and the high-backed, old-fashioned arm-chair, which she had occupied, remained vacant at the opposite corner of the fire-place. Puss, who lay with her head on my arm, became more and more sleepy, and I pondered on the propriety of preparing for bed.

"Of a sudden I became aware that something had affected my pet's equanimity. The purring ceased, and she exhibited rapidly increasing symptoms of uneasiness. I bent down, and endeavored to coax her into quietness; but she instantly struggled to her feet in my lap, and, spitting vehemently, with back arched and tail swollen, she assumed a mingled attitude of terror and defiance.

"The change in her position obliged me to raise my head; and on looking up, to my inexpressible horror, I then perceived that a little, hideous, wrinkled old hag occupied mamma's chair. Her hands were rested on her knees, and her body was stooped forward so as to bring her face in close proximity with mine. Her eyes, piercingly fierce and shining with an overpowering lustre, were steadfastly fixed on me. It was as if a fiend were glaring at me through them. Her dress and general appearance denoted her to belong to the French *bourgeoisie;* but those eyes, so wonderfully large, and in their expression so intensely wicked, entirely absorbed my senses, and precluded any attention to detail. I should have screamed, but my breath was gone while that terrible gaze so horribly fascinated me: I could neither withdraw my eyes nor rise from my seat.

"I had meanwhile been trying to keep a tight hold on the cat, but she seemed resolutely determined not to remain in such ugly neighborhood, and, after some most desperate efforts, at length succeeded in escaping from my grasp. Leaping over tables, chairs, and all that came in her way, she repeatedly threw herself, with frightful violence, against the top panel of the door which communicated with the disused room. Then, returning in the same frantic manner, she furiously dashed against the door on the opposite side.

"My terror was divided, and I looked by turns, now at the old woman, whose great staring eyes were constantly fixed on me, and now at the cat, who was becoming every instant more frantic. At last the dreadful idea that the animal had gone mad had the effect of restoring my breath, and I screamed loudly.

"Mamma ran in immediately, and the cat, on the door opening, literally sprang over her head, and for upward of half an hour ran up and down stairs as if pursued. I turned to point to the object of my terror: it was gone. Under such circumstances the lapse of time is difficult to appreciate, but I should think that the apparition lasted about four or five minutes.

"Some time afterward it transpired that a former proprietor of the house, a woman, had hanged herself in that very room."

The close but evidently unsuspected resemblance of this narrative to the story of Balaam is worthy of notice. In both cases we have the remarkable fact that the animal was the first to see the spiritual being, and to show by its terrified actions that it had done so.

There are but a few words to be said by way of epilogue.

Some of the objections that have been made to the future life of the lower animals have already been mentioned, but there are two others which I must briefly notice. One is that, if all created beings are to live eternally in heaven, there would not be room for them. I feel almost ashamed even to mention such an absurd notion, but as it has been put forward by several persons I feel bound to notice it.

The answer is self-evident. In the first place, in the spiritual world space and time do not exist; and even if they did, surely God can create space, if he has need of it.

The second objection is that by granting immortality to the animals we lower the condition of humanity; but if the animals be immortal there is surely no use in denying it. We can not shirk a fact, and even if we could we ought not to do so. Such an argument, moreover, is not very creditable to humanity, for it seeks to elevate man by depreciating his fellow-creatures of a lower order.

In announcing my belief that the lower animals share immortality with man in the next world, as they share mortality in this, I do not claim for them the slightest equality. Man will be man, and beast will be beast, and insect will be insect, in the next world as in this. They are living exponents of divine ideas, as is evident from the Holy Scriptures, and will be wanted to continue in the world of spirit the work which they have begun in the world of matter.

But, though I do not claim for them the slightest equality with man, I do claim for them a higher status in creation than is generally attributed to them; I do claim for them a future life, in which they can be compensated for the sufferings which so many of them have to undergo in this world; and I do so chiefly because I am quite sure that most of the cruelties which are perpetrated on the animals are due to the habit of considering them as mere machines, without susceptibilities, without reason, and without the capacity of a future.

THE END.

Printed in Dunstable, United Kingdom